Dissertation zur Erlangung des Doktorgrades
der Fakultät für Chemie und Pharmazie
der Ludwig–Maximilians–Universität München

Relativistic electronic transport theory -
The spin Hall effect
and related phenomena

Stephan Lowitzer

aus

Dachau, Deutschland

2010

Bibliografische Information der Deutschen Nationalbibliothek

Die Deutsche Nationalbibliothek verzeichnet diese Publikation in der
Deutschen Nationalbibliografie; detaillierte bibliografische Daten sind
im Internet über http://dnb.d-nb.de abrufbar.

ISBN 978-3-8325-2629-0

Logos Verlag Berlin GmbH
Comeniushof, Gubener Str. 47,
10243 Berlin
Tel.: +49 (0)30 42 85 10 90
Fax: +49 (0)30 42 85 10 92
INTERNET: http://www.logos-verlag.de

To Maxi

Contents

Contents

Chapter 1

Introduction

In principle the determination of the electric conductivity (or resistivity) of a conducting perfect crystal is a "simple" problem. An electron which moves in a periodic potential can propagate without any effective scattering due to the fact that the coherent scattered waves interfere constructively [1, 2]. This leads to an infinite conductivity or equivalently the resistivity becomes zero. This is in contrast with experimental observations which show that the resistivity of metals is non-zero. The reason for this discrepancy is that in real solids the periodicity of the potential is distorted due to various phenomena like thermally induced atomic displacements, lattice distortions (e.g. screw and edge dislocations) or atomic and magnetic disorder [3]. The aim of the present work is to investigate on an *ab initio* level transport phenomena as the residual resistivity of alloys where the ideal lattice periodicity is distorted due to disorder of the atomic lattice site occupation.

Due to the fact that even very low impurity concentrations can drastically influence transport phenomena it is obvious that for technical applications within standard electronics or spintronics (see below) it is crucial to understand the underlying mechanisms which are responsible for the modification of transport properties of a certain material.

During the last years a new research area emerged which is called spintronics [4–8]. Spintronics is a technology that exploits the intrinsic spin of the electron and its associated magnetic moment in addition to its fundamental electronic charge. The central issue of this multidisciplinary field is the manipulation of the spin degree of freedom in solid-state systems [6]. One of the most prominent effects which belongs to the field of spintronics is the giant magnetoresistance (GMR) effect. The GMR effect was discovered independently by A. Fert [9] and P. Grünberg [10] for which they have been awarded the 2007 Nobel Prize in Physics. A typical GMR device consists of two magnetic layers which are separated by an additional non-magnetic layer (e.g. Co/Cu/Co). If one measures the resistivity of such a device one obtains a strong dependence of the resistivity on the relative orientation of the magnetic configuration of the two Fe layers. A ferromagnetic alignment of the Fe layers lead to a different resistivity as compared to an anti-ferromagnetic configuration. The industrial importance of this effect is demonstrated by the fact that nowadays the GMR is widely used in read heads of modern hard drives [11].

Due to the fact that spintronic devices generically need an imbalance between spin-up and spin-down populations of electrons [8] it seems almost a given fact that ferromagnetic components are necessary for the construction of spintronic devices. Discoveries in recent years have inspired a completely different route in spintronic research which need no ferromagnetic components [8]. The research field "spintronic without magnetism" is based on the possibility to manipulate electric currents via spin-orbit coupling only. Spin-orbit coupling generates spin-polarization and therefore allows the generation and manipulation of spins solely by electric fields. The advantage of "spintronics without magnetism" compared to standard spintronics is the reduced device complexity which is considerable in standard spintronic devices due to the incorporation of local magnetic fields into the device architecture [8]. Generating and manipulating the spin polarization is one of the important prerequisites for the realization of new spintronic devices [6]. The spin Hall effect (SHE) is considered as a convenient method for generating spin polarization, in addition to traditional methods like spin injection from ferromagnetic metals [12]. The SHE appears when an electric current flows through a medium with spin-orbit coupling present, leading to a spin-current perpendicular to the charge current. This effect is even present in non-magnetic materials as could be demonstrated experimentally e.g. for Pt [13]. The SHE was first described 1971 by Dyakonov and Perel [14, 15] and more recently by Hirsch [16]. This effect is illustrated schematically in Fig. 1.1. The electric current splits into a spin-up and spin-down part which leads to a spin accumulation at the edges without any accompanying Hall voltage. In summary, due to the fact that establishing techniques for efficient generation and manipulation of spin-currents is a key for further advancement of spintronic devices the SHE can be considered as one of the most promising effects in recent spintronic research [13]. Therefore, the electrical conductivity tensor which includes the spin Hall conductivity coefficient is one of the central quantities within spintronics. The main issue of the present work is to study in detail the underlying mechanisms of the aforementioned effects.

The present work is organized as follows: In chapter 2 the fundamentals of density functional theory (DFT) are presented which are the basis of the

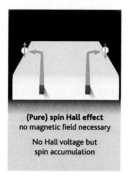

Figure 1.1: Schematic picture of the spin Hall effect [17].

calculations shown in this work. In chapter 3 the relativistic Korringa-Kohn-Rostoker Green's function method (KKR-GF) is discussed. This method has been used for the calculation of the electronic structure of the investigated systems. In order to describe the electronic structure of alloys the coherent potential approximation (CPA) as well as the non-local coherent potential approximation (NLCPA) are introduced. Chapter 4 discusses in detail the derivation of the Kubo equation which can be used for the *ab initio* calculation of the conductivity tensor within linear response theory. Starting from the Kubo equation the Kubo-Středa equation is derived using the independent electron approximation. This equation is the basis of all transport calculations of the present work. In chapter 5 the spin decomposition problem within a fully relativistic description is addressed. Due to the fact that the well known non-relativistic spin operator does not commute with the Dirac equation the spin is no longer conserved and not a good quantum number if spin-orbit coupling is present. Therefore, using relativistic polarization operators spin projection operators are derived and applied to several alloy systems. The results are compared to an approximate spin decomposition scheme. In chapter 6 the the residual resistivity of the diluted magnetic semiconductor system $Ga_{1-x}Mn_xAs$ is discussed. In addition, the influence of short ranged correlations of the atomic lattice site occupation is investigated using the NLCPA. It is demonstrated that the transport formalism using the NLCPA is able to describe the counterintuitive K-effect which is connected with an increase of the residual resistivity with increasing ordering. Chapter 7 investigates the SHE for several non-magnetic 4d- and 5d-transition metal alloys. The anomalous Hall effect (AHE) is also discussed for magnetic Pd based alloys with $3d$-transition metals. The SHE as well as AHE is decomposed into intrinsic and extrinsic (skew and side-jump scattering) contributions. In order to complete the Hall effect discussion the orbital Hall effect is addressed in addition.

In summary, the major aim of the present work is to investigate several transport phenomena like e.g. residual resistivity, spin Hall effect as well as anomalous Hall effect of alloys on a fully relativistic *ab initio* level.

Chapter 2

Density Functional Theory

Density functional theory (DFT) is a sophisticated way to substitute an extremely complicated many-body problem by an effective one-body problem. The central quantity within DFT is the electron density $\rho(\mathbf{r})$. DFT allows one to work with the electron density $\rho(\mathbf{r})$ instead of the complicated N-electron wave function $\Psi(\mathbf{x}_1, \mathbf{x}_2, ..., \mathbf{x}_N)$.

The first scheme which used $\rho(\mathbf{r})$ as a central quantity like "modern" DFT have been performed by Thomas [18] and Fermi [19]. 1964 Hohenberg and Kohn [20] published a paper in which they proved that there is a one-to-one mapping between the ground state wave function and the ground state electron density ($\Psi_0 \Leftrightarrow \rho_0$) and also between the potential v_{ext} and the ground state wave function ($v_{\text{ext}} \Leftrightarrow \Psi_0$). The combination of these two mappings show that v_{ext}, Ψ_0 and ρ_0 determine each other mutually and uniquely ($v_{\text{ext}} \Leftrightarrow \Psi_0 \Leftrightarrow \rho_0$). The important observation that the ground state is a unique functional of the ground state density $|\psi[\rho_0]\rangle$ implies that the density determines all electronic ground state properties of the investigated system [21].

2.1 Density Variational Principle

Hohenberg and Kohn [20] proofed their observations via a *reductio ad absurdum* procedure. Levy [22] showed that it is also possible to use a *constrained search* approach to proof that the external potential is uniquely determined by the ground state density. In the following the proof by Levy [22] is sketched.

The variational principle states that the ground state energy E can be found by a minimizing procedure

$$E = \min_{\Psi} \langle \Psi | \hat{H} | \Psi \rangle \tag{2.1}$$

over all normalized and antisymmetric wave functions $|\Psi\rangle$ [23]. $\hat{H}$ describes the fully interacting many-body system:

$$\hat{H} = \hat{T} + \hat{V}_{ee} + \hat{V} \, , \qquad (2.2)$$

with the kinetic energy operator $\hat{T}$, the electron electron interaction operator $\hat{V}_{ee}$ and the external potential operator $\hat{V}$. The next step is to split the minimizing procedure into two parts. The first part is a minimization over all $|\Psi\rangle$ which yield a given density $\rho(\mathbf{r})$ and in the second part one has to minimize over all N-electron densities $\rho(\mathbf{r})$:

$$E = \min_{\rho} \left[\underbrace{\min_{\Psi \to \rho} \langle \Psi | \hat{T} + \hat{V}_{ee} | \Psi \rangle + \int d^3r \, v_{\text{ext}}(\mathbf{r})\rho(\mathbf{r})}_{=F[\rho]} \right] . \qquad (2.3)$$

In order to take into account that only N-electron densities are considered one introduces a Lagrange multiplier μ into the functional variation:

$$\frac{\delta}{\delta\rho(\mathbf{r})} \left\{ F[\rho] + \int d^3r \, v_{\text{ext}}(\mathbf{r})\rho(\mathbf{r}) + \mu \left(N - \int d^3r \, \rho(\mathbf{r}) \right) \right\} = 0 \, , \qquad (2.4)$$

which is equivalent to the Euler equation:

$$\frac{\delta F}{\delta\rho(\mathbf{r})} + v_{\text{ext}}(\mathbf{r}) = \mu \, . \qquad (2.5)$$

Eq. (2.5) shows that the external potential $v_{\text{ext}}(\mathbf{r})$ is uniquely determined by the ground state density [24].

2.2 Kohn-Sham Equation

The application of the equations shown above is hindered due to the fact that it is not clear how to express $T[\rho]$ and $V_{ee}[\rho]$ as functionals of the density ρ. The traditional Thomas-Fermi model or the improved Thomas-Fermi-Dirac model [25] are based on drastic assumptions which lead to a restricted applicability of these models. 1965 Kohn and Sham [26] achieved a break through concerning the treatment of the kinetic energy functional $T[\rho]$. They postulated the existence of a non-interacting reference system which has the same ground state density ρ_0 as the interacting system.

For a non-interacting system of electrons Eq. (2.5) transforms to:

$$\frac{\delta T_s}{\delta\rho(\mathbf{r})} + v_s(\mathbf{r}) = \mu \, , \qquad (2.6)$$

with the kinetic energy of a non-interacting system $T_s[\rho]$ and the Kohn-Sham potential v_s. The interacting system is connected with the non-interacting system via the equation:

$$F[\rho] = T_s[\rho] + U[\rho] + E_{xc}[\rho] \ . \qquad (2.7)$$

$U[\rho]$ is the classical part of $V_{ee}[\rho]$ which describes the Coulomb potential energy:

$$U[\rho] = \frac{1}{2} \frac{e^2}{4\pi\epsilon_0} \int \int d^3r \, d^3r' \frac{\rho(\mathbf{r})\rho(\mathbf{r}')}{|\mathbf{r}-\mathbf{r}'|} \ . \qquad (2.8)$$

The exchange-correlation energy $E_{xc}[\rho]$ is defined via Eq. (2.7) and therefore take care that the the the Euler Eqs. (2.5) and (2.6) are consistent with each other implying the relation:

$$v_s(\mathbf{r}) = v_{\text{ext}}(\mathbf{r}) + \frac{\delta U[\rho]}{\delta\rho(\mathbf{r})} + \frac{\delta E_{xc}[\rho]}{\delta\rho(\mathbf{r})} \ . \qquad (2.9)$$

At a first view there is no advance through the change from the interacting to the non-interacting system because it is not clear how to calculate $E_{xc}[\rho]$. Fortunately, it turns out that $T_s[\rho]$ typically captures a very large part of the energy whereas $E_{xc}[\rho]$ is a smaller part [24]. Therefore, $E_{xc}[\rho]$ is much more appropriate to be treated in an approximate way.

In order to continue with the derivation of the Kohn-Sham equations one has to consider that the non-interacting N-particle many-body Schrödinger equation can be reduced by separation of the variables to N single particle equations. From the solutions of the single particle equations one can easily construct the N particle wave function with the help of a Slater determinant of Kohn-Sham orbitals ϕ_i. These observations lead to the non-relativistic Kohn-Sham equations:

$$\left(-\frac{\hbar^2}{2m} \nabla^2 + v_{\text{ext}}(\mathbf{r}) + \frac{e^2}{4\pi\epsilon_0} \int d^3r' \frac{\rho(\mathbf{r}')}{|\mathbf{r}-\mathbf{r}'|} + \frac{\delta E_{xc}[\rho]}{\delta\rho(\mathbf{r})} \right) \phi_i(\mathbf{r}) = \epsilon_i \phi_i(\mathbf{r}) \ , \qquad (2.10)$$

with

$$\frac{\delta U[\rho]}{\delta\rho(\mathbf{r})} = \frac{e^2}{4\pi\epsilon_0} \int d^3r' \frac{\rho(\mathbf{r}')}{|\mathbf{r}-\mathbf{r}'|} \qquad (2.11)$$

$$\rho(\mathbf{r}) = \sum_{i=1}^{N} \phi_i^\dagger(\mathbf{r})\phi_i(\mathbf{r}) \ . \qquad (2.12)$$

2.2.1 Relativistic Formulation

The relativistic Hohenberg-Kohn theorem was first formulated by Rajagopal and Callaway [27] and later amplified by Ramana and Rajagopal [28]. It states that the ground-state energy is a unique functional of the ground-state four-current $J^\mu(\mathbf{r})$. Therefore, the proper name of this theory is four-current density functional theory [29]. Within this theory one obtains the Dirac-Kohn-Sham equations [30]:

$$\left[c\boldsymbol{\alpha} \cdot (\hat{\mathbf{p}} - e\mathbf{A}_{\text{eff}}(\mathbf{r})) + \beta mc^2 + v_{\text{eff}}(\mathbf{r}) \right] \phi_i(\mathbf{r}) = \epsilon_i \phi_i(\mathbf{r}) \,, \qquad (2.13)$$

with

$$v_{\text{eff}}(\mathbf{r}) = v_{\text{ext}}(\mathbf{r}) + \frac{e^2}{4\pi\epsilon_0} \int d^3 r' \frac{J^0(\mathbf{r}')}{|\mathbf{r} - \mathbf{r}'|} + \frac{\delta E_{xc}[J^\mu(\mathbf{r})]}{\delta J^0(\mathbf{r})} \qquad (2.14)$$

$$\mathbf{A}_{\text{eff}}(\mathbf{r}) = \mathbf{A}_{\text{ext}}(\mathbf{r}) + \frac{e}{4\pi\epsilon_0 c^2} \int d^3 r' \frac{\mathbf{J}(\mathbf{r}')}{|\mathbf{r} - \mathbf{r}'|} + \frac{1}{e}\frac{\delta E_{xc}[J^\mu(\mathbf{r})]}{\delta \mathbf{J}(\mathbf{r})} \qquad (2.15)$$

$$J^0(\mathbf{r}) = \sum_{i=1}^{N} \phi_i^\dagger(\mathbf{r})\phi_i(\mathbf{r}) \qquad\qquad \mathbf{J}(\mathbf{r}) = c\sum_{i=1}^{N} \phi_i^\dagger(\mathbf{r})\boldsymbol{\alpha}\phi_i(\mathbf{r}) \qquad (2.16)$$

and with the standard 4×4 Dirac matrices α_i, β [31]. According to the fact that the Dirac-Kohn-Sham equation contains 4×4 matrices the wave functions are four spinors.

In order to make applications of Eq. (2.13) feasible the four-current density $J^\mu(\mathbf{r})$ can be split by a Gordon decomposition. This procedure decomposes $J^\mu(\mathbf{r})$ into terms which have a clear physical interpretation. In the stationary situation the three-current density $\mathbf{J}(\mathbf{r})$ can be written as a sum of orbital and spin-current densities [32]. If one completely neglects the orbital current and considers only the spin magnetization density $\mathbf{m}(\mathbf{r})$ which correspond to a collinear alignment of the spins (the orientation of $\mathbf{m}(\mathbf{r})$ is globally fixed as $m(\mathbf{r})\mathbf{e}_z$) [33] Eq. (2.13) simplifies drastically [34]:

$$\left[c\boldsymbol{\alpha} \cdot \hat{\mathbf{p}} + \beta mc^2 + v_{\text{eff}}(\mathbf{r}) + \beta\Sigma_z B_{\text{eff}}(\mathbf{r}) \right] \phi_i(\mathbf{r}) = \epsilon_i \phi_i(\mathbf{r}) \qquad (2.17)$$

with

$$v_{\text{eff}}(\mathbf{r}) = v_{\text{ext}}(\mathbf{r}) + \frac{e^2}{4\pi\epsilon_0} \int d^3 r' \frac{\rho(\mathbf{r}')}{|\mathbf{r} - \mathbf{r}'|} + \frac{\delta E_{xc}[\rho(\mathbf{r}), m(\mathbf{r})]}{\delta\rho(\mathbf{r})} \qquad (2.18)$$

$$B_{\text{eff}}(\mathbf{r}) = B_{\text{ext}}(\mathbf{r}) + \frac{\delta E_{xc}[\rho(\mathbf{r}), m(\mathbf{r})]}{\delta m(\mathbf{r})} \qquad (2.19)$$

$$m(\mathbf{r}) = \sum_{i=1}^{N} \phi_i^\dagger(\mathbf{r})\beta\Sigma_z\phi_i(\mathbf{r}) \qquad\qquad \Sigma_z = \sigma_z \otimes \mathbb{1}_2 \,, \qquad (2.20)$$

where $B_{ext}(\mathbf{r})$ is an external magnetic field. Eq. (2.17) is widely used in computations and is also used in the present work with the restriction that the potential terms are spherically symmetric (the atomic sphere approximation (ASA) has been used):

$$v_{\text{eff}}(\mathbf{r}) = v_{\text{eff}}(r) \ . \tag{2.21}$$

2.2.2 The Exchange-Correlation Energy

Density functional theory is in principle an exact theory. The major problem of DFT is that no exact expression for the exchange-correlation energy functional $E_{xc}[\rho(\mathbf{r})]$ is available. Nevertheless, since the famous paper of Hohenberg and Kohn [20] much work has been done to develop reliable approximations for $E_{xc}[\rho(\mathbf{r})]$ [35]. Although, $E_{xc}[\rho(\mathbf{r})]$ often is a small fraction of the total energy, $E_{xc}[\rho(\mathbf{r})]$ is a kind of "glue" that allows atoms to build bonds [21]. Thus, accurate approximations for $E_{xc}[\rho(\mathbf{r})]$ are essential for meaningful calculations.

The most simple approximation for $E_{xc}[\rho(\mathbf{r})]$ is the local density approximation (LDA). In this approximation, electronic properties are determined as functionals of the electron density by applying locally relations appropriate for a homogeneous electronic system [21]. In order to calculate spin polarized systems Kohn and Sham [26] proposed the local spin density approximation (LSDA):

$$E_{xc}^{\text{LSDA}}[\rho_\uparrow(\mathbf{r}), \rho_\downarrow(\mathbf{r})] = \int d^3 r \rho(\mathbf{r}) e_{xc}[\rho_\uparrow(\mathbf{r}), \rho_\downarrow(\mathbf{r})] \ , \tag{2.22}$$

with $\rho(\mathbf{r}) = \rho_\uparrow(\mathbf{r}) + \rho_\downarrow(\mathbf{r})$. The quantity $e_{xc}[\rho_\uparrow(\mathbf{r}), \rho_\downarrow(\mathbf{r})]$ is the known [36] exchange-correlation energy per particle for an electron gas of uniform spin densities $\rho_\uparrow(\mathbf{r})$ and $\rho_\downarrow(\mathbf{r})$, respectively. The LSDA is the most popular method in solid state physics for electronic structure calculations whereas the generalized gradient approximation (GGA) is more common in quantum chemistry [24]. The GGA is an extension of the LDA which takes into account the gradient of the density:

$$E_{xc}^{\text{GGA}}[\rho_\uparrow(\mathbf{r}), \rho_\downarrow(\mathbf{r})] = \int d^3 r f[\rho_\uparrow(\mathbf{r}), \rho_\downarrow(\mathbf{r}), \nabla \rho_\uparrow(\mathbf{r}), \nabla \rho_\downarrow(\mathbf{r})] \ . \tag{2.23}$$

Due to the fact that the gradient of the density in real materials becomes very large a simple expansion of the density breaks down. Therefore, many different ways were proposed which modify the behavior at large gradients. While there is only one type of LDA there are several different ways to account for density gradients of the GGA like e.g. B88 [37], PW91 [38] and PBE [39].

9

A very frequently used functional in the chemistry community is the so-called B3LYP functional [35], that belongs to the class of hybrid functionals. Hybrid functionals are defined as functionals which consists of a mixture of exact exchange from Hartree-Fock theory in combination with exchange and correlation from LDA or GGA.

In relativistic DFT the construction of $E_{xc}[\rho(\mathbf{r}), m(\mathbf{r})]$ becomes even more involved due to the fact that nothing is known about the $m(\mathbf{r})$ dependence of $E_c[\rho(\mathbf{r}), m(\mathbf{r})]$ up to know [33]. Therefore, relativistic DFT is usually applied in conjunction with non-relativistic spin-density functionals $E_{xc}^{\mathrm{nrel}}[\rho_\uparrow(\mathbf{r}), \rho_\downarrow(\mathbf{r})]$ which are adapted in the following way [33]:

$$\rho_\pm(\mathbf{r}) = \frac{1}{2}\left[\rho(\mathbf{r}) \mp \frac{1}{\mu_B}m(\mathbf{r})\right] \tag{2.24}$$

$$\frac{\delta E_{xc}[\rho(\mathbf{r}), m(\mathbf{r})]}{\delta\rho(\mathbf{r})} = \frac{1}{2}\left\{\frac{\delta E_{xc}^{\mathrm{nrel}}[\rho_+(\mathbf{r}), \rho_-(\mathbf{r})]}{\delta\rho_+(\mathbf{r})} + \frac{\delta E_{xc}^{\mathrm{nrel}}[\rho_+(\mathbf{r}), \rho_-(\mathbf{r})]}{\delta\rho_-(\mathbf{r})}\right\} \tag{2.25}$$

$$\frac{\delta E_{xc}[\rho(\mathbf{r}), m(\mathbf{r})]}{\delta m(\mathbf{r})} = \frac{1}{2}\left\{\frac{\delta E_{xc}^{\mathrm{nrel}}[\rho_+(\mathbf{r}), \rho_-(\mathbf{r})]}{\delta\rho_-(\mathbf{r})} - \frac{\delta E_{xc}^{\mathrm{nrel}}[\rho_+(\mathbf{r}), \rho_-(\mathbf{r})]}{\delta\rho_+(\mathbf{r})}\right\}. \tag{2.26}$$

Chapter 3

Multiple Scattering Theory

Multiple scattering theory (MST) was first used in 1892 by Lord Rayleigh [40] to describe the propagation of heat or electricity through inhomogeneous media. The power of this method is illustrated by the fact that a medium which consists of several non-overlapping potentials can be treated in such a way that the single potentials are independently investigated and afterwards the observations are combined to describe the complete medium [41]. This is an enormous advantage because a N-potential problem can be decoupled to N single potential problems.

In the present work MST has been employed for the calculation of the electronic structure of solid materials. 1947 Korringa [42] demonstrated that MST can be used to calculate eigenvalues and eigenvectors associated with the electronic states of a periodic system via a secular equation. An equivalent equation was found by Kohn and Rostoker [43] with the help of a variational formalism. The method which evolves out of these pioneering works together with several other works e.g. [44–48] is nowadays called the Korringa-Kohn-Rostoker (KKR) method or the KKR Green's function method (KKR-GF)(accordingly for the spin-polarized relativistic case: SPR-KKR-GF). A characteristic and remarkable feature of the KKR-GF method is the complete separation of the structural aspects (the geometrical arrangement of the scattering potentials) from potential aspects. The structural aspects of the investigated material can be independently calculated which leads to the so-called structure constants for a certain lattice. The whole potential aspects are contained in the t-matrix which has to be determined for each potential type. The decoupling of the potential aspects from the structural aspects allows a computational efficient application of the KKR-GF method.

The phrase "Green's function" in the acronym KKR-GF indicates that this method gives in a naturally way access to the Green's function of the investigated system. The access to the Green's function is an important advantage compared to other methods for the determination of the electronic structure due to the fact that many solid state physic problems are theoretical described via the Green's function as for example transport and spectroscopic properties [49, 50]. Another advantage of the KKR-GF method is the fact that disordered systems can be treated in a very accurate way with the help of the coherent potential approximation (CPA). The CPA is a mean field theory for the electronic structure of disordered alloys and has to be seen as the best single-site solution to this problem [41]. However, alloys often exhibit short-ranged correlation effects in the lattice site occupation which leads to a locally ordered structure. Even such effects like short-ranged ordering effects can be described on an *ab initio* level via the KKR-GF method in conjunction with the non-local coherent potential approximation (NLCPA).

The combination of all of the above mentioned properties make the KKR-GF method to a very powerful and flexible tool for the investigation of various solid state phenomena.

3.1 Single-Site Scattering

3.1.1 The Dirac Equation for Free Electrons

The Dirac equation without a potential can be written as [31]:

$$(c\boldsymbol{\alpha} \cdot \hat{\mathbf{p}} + \beta mc^2)\psi(\mathbf{r}) = E\psi(\mathbf{r}) , \tag{3.1}$$

with $\hat{\mathbf{p}} = -i\hbar\nabla$ and the Dirac matrices:

$$\boldsymbol{\alpha} = \begin{pmatrix} 0 & \boldsymbol{\sigma} \\ \boldsymbol{\sigma} & 0 \end{pmatrix} \qquad \text{and} \qquad \beta = \begin{pmatrix} \mathbb{1}_2 & 0 \\ 0 & -\mathbb{1}_2 \end{pmatrix} . \tag{3.2}$$

In Eq. (3.2) $\boldsymbol{\sigma}$ denotes the Pauli matrices:

$$\sigma_x = \begin{pmatrix} 0 & 1 \\ 1 & 0 \end{pmatrix} \qquad \sigma_y = \begin{pmatrix} 0 & -i \\ i & 0 \end{pmatrix} \qquad \sigma_z = \begin{pmatrix} 1 & 0 \\ 0 & 1 \end{pmatrix} \tag{3.3}$$

and $\mathbb{1}_2$ denotes the 2×2 unity matrix. $\boldsymbol{\alpha}$ as well as β are 4×4 matrices and ψ is a bi-spinor. If one compares the properties of the Schrödinger equation with the properties of the Dirac equation one finds that the Dirac Hamiltonian H_D no longer commutes with $\hat{l}^2$, $\hat{l}_z$ and σ_z. This clearly indicates that in a relativistic theory l, m_l as well as m_s are no longer good quantum numbers. It turns out that H_D commutes with the operators $\boldsymbol{\sigma}^2$, $\hat{\boldsymbol{j}}^2$ and $\hat{j}_z$ with j the total angular momentum and additional with the spin-orbit operator

$\hat{K} = \beta(\mathbb{1}_4 + \hat{\boldsymbol{l}} \cdot \boldsymbol{\sigma})$. A set of eigenfunctions to the operators $\{\boldsymbol{\sigma}^2, \hat{\boldsymbol{j}}^2, \hat{\boldsymbol{j}}_z, \hat{K}\}$ are the spin-angular functions [31]:

$$\chi_\Lambda(\hat{\mathbf{r}}) = \sum_{m_s} C(l\tfrac{1}{2}j; \mu - m_s, m_s) Y_l^{\mu - m_s}(\hat{\mathbf{r}}) \chi_{m_s} , \qquad (3.4)$$

with the Clebsch-Gordan coefficients $C(l\tfrac{1}{2}j; \mu - m_s, m_s)$, the spherical harmonics $Y_l^{\mu - m_s}$ and the spin eigenvectors $\chi_{1/2}^T = (1, 0)$, $\chi_{-1/2}^T = (0, 1)$ ($\hat{\mathbf{r}}$ denotes the angular dependent part of $\mathbf{r}$). The eigenvalue relations are:

$$\boldsymbol{\sigma}^2 \, \chi_\Lambda(\hat{\mathbf{r}}) = \frac{3}{4} \, \chi_\Lambda(\hat{\mathbf{r}}) \qquad (3.5)$$

$$\hat{\boldsymbol{j}}^2 \, \chi_\Lambda(\hat{\mathbf{r}}) = j(j+1) \, \chi_\Lambda(\hat{\mathbf{r}}) \qquad (3.6)$$

$$\hat{\boldsymbol{j}}_z \, \chi_\Lambda(\hat{\mathbf{r}}) = \mu \, \chi_\Lambda(\hat{\mathbf{r}}) \qquad (3.7)$$

$$\hat{K} \, \chi_\Lambda(\hat{\mathbf{r}}) = -\kappa \, \chi_\Lambda(\hat{\mathbf{r}}) , \qquad (3.8)$$

with $\Lambda = (\kappa, \mu)$.

In order to solve Eq. (3.1) one can use the following ansatz [31]:

$$\psi_\kappa^\mu(\mathbf{r}) = \begin{pmatrix} g_\kappa(r) \chi_\kappa^\mu(\hat{\mathbf{r}}) \\ i f_\kappa(r) \chi_{-\kappa}^\mu(\hat{\mathbf{r}}) \end{pmatrix} . \qquad (3.9)$$

Now one can substitute this bi-spinor ansatz into the Dirac equation in polar coordinates and obtains the following expressions for the radial wave functions [31]:

$$\frac{\partial^2 P_\kappa(r)}{\partial r^2} = \left(\frac{\kappa(\kappa+1)}{r^2} - \frac{p^2}{\hbar^2} \right) \qquad (3.10)$$

$$Q_\kappa(r) = c^2 \hbar (E + mc^2)^{-1} \left(\frac{\partial P_\kappa(r)}{\partial r} + \frac{\kappa}{r} P_\kappa(r) \right) , \qquad (3.11)$$

with $P_\kappa(r) = rg_\kappa(r)$, $Q_\kappa(r) = crf_\kappa(r)$ and p^2 is defined via the relativistic energy-momentum relation $E^2 = p^2 c^2 + m^2 c^4$. Eq. (3.10) can be transformed into a differential equation of the general form:

$$x^2 \frac{d^2 R}{dx^2} + 2x \frac{dR}{dx} + [x^2 - n(n+1)]R = 0 \qquad (3.12)$$

which is the spherical Bessel differential equation [51]. The solutions to this differential equation are the spherical Bessel functions $j_l(pr)$, the spherical Neumann functions $n_l(pr)$ and the spherical Hankel functions:

$$h_l^\pm(pr) = j_l(pr) \pm i n_l(pr) , \qquad (3.13)$$

13

with $p = \sqrt{\frac{E^2 - m^2 c^4}{c^2}}$. Finally, one arrives at the following expression for the most general eigenfunctions of the free Dirac equation in spherical coordinated [52]:

$$\psi_\kappa^\mu(\mathbf{r}) = \begin{pmatrix} [\cos \delta_\kappa \, j_l(pr) - \sin \delta_\kappa \, n_l(pr)] \chi_\kappa^\mu(\hat{\mathbf{r}}) \\ \frac{ipcS_\kappa}{E + mc^2} [\cos \delta_\kappa \, j_{\bar{l}}(pr) - \sin \delta_\kappa \, n_{\bar{l}}(pr)] \chi_{-\kappa}^\mu(\hat{\mathbf{r}}) \end{pmatrix} , \qquad (3.14)$$

with $S_\kappa = \operatorname{sgn} \kappa$, $\bar{l} = l - S_\kappa$ and $\cos \delta_\kappa$ as well as $\sin \delta_\kappa$ have to be considered as coefficients of the Bessel and Neumann functions. Restricting to physically acceptable solutions for the free electron case that are regular at the origin one ends up with:

$$\psi_\kappa^\mu(\mathbf{r}) = \begin{pmatrix} j_l(pr) \chi_\kappa^\mu(\hat{\mathbf{r}}) \\ \frac{ipcS_\kappa}{E + mc^2} \, j_{\bar{l}}(pr) \chi_{-\kappa}^\mu(\hat{\mathbf{r}}) \end{pmatrix} . \qquad (3.15)$$

This section shows that in the case of a relativistic free-particle the Dirac equation can be decomposed into a radial and a angular dependent part. The radial dependent part leads as a corresponding non-relativistic quantum mechanical treatment of a free-particle to the well known spherical Bessel differential equation with the spherical Bessel, Neumann and Hankel functions as solutions. An important difference to the non-relativistic theory is that the solutions of the spherical Bessel equation contain as argument the relativistic k which is defined via the relativistic energy-momentum relation. The ambiguity in the choice of the sign of the square root $\sqrt{\frac{E^2 - m^2 c^4}{c^2}}$ leads to the particle anti-particle problem in a relativistic theory [31]. In the present work only states corresponding to positive energies are considered and therefore interactions between the negative energy continuum states and the positive energy states are neglected.

3.1.2 The Relativistic Free-Particle Green's Function

In general the Green's function can be defined as the solution of the inhomogeneous differential equation [50]:

$$(E - \hat{H}) G(\mathbf{r}, \mathbf{r}', E) = \delta(\mathbf{r} - \mathbf{r}') , \qquad (3.16)$$

with an arbitrary Hamiltonian $\hat{H}$ which contains a differential operator. A formal solution to Eq. (3.16) is:

$$G(\mathbf{r}, \mathbf{r}', E) = \sum_n \frac{\langle \mathbf{r} | \phi_n \rangle \langle \phi_n | \mathbf{r} \rangle}{E - \lambda_n} , \qquad (3.17)$$

with

$$\hat{H}|\phi_n\rangle = \lambda_n|\phi_n\rangle . \tag{3.18}$$

If $\hat{H}$ is a Hermitian operator the Green's function shows singularities only at the real eigenvalues of $\hat{H}$. To overcome these singularities one can transfer the Green's function into the complex plane where it becomes an analytic function. This transferring into the complex plane can be done with the help of a limiting procedure:

$$G^{\pm} = \lim_{\eta \to 0^+} (E - \hat{H} \pm \eta)^{-1} . \tag{3.19}$$

G^+ and G^- are called the retarded and advanced Green's functions, respectively. In the following the superscript is omitted implying that always the retarded Green's function is meant.

The relativistic free-particle Green's function is defined via the following equation [52]:

$$(E - c\boldsymbol{\alpha} \cdot \hat{\mathbf{p}} - \beta mc^2)G_0(\mathbf{r}, \mathbf{r}', E) = \delta(\mathbf{r} - \mathbf{r}')\mathbb{1}_4 , \tag{3.20}$$

which can be transformed to:

$$G_0(\mathbf{r}, \mathbf{r}', E) = -\frac{1}{\hbar^2 c^2}(c\boldsymbol{\alpha} \cdot \hat{\mathbf{p}} + \beta mc^2 + E)\frac{e^{ipR}}{4\pi \mathbf{R}} \tag{3.21}$$

($\mathbf{R} = \mathbf{r} - \mathbf{r}'$). With the help of the identity [31]:

$$\frac{e^{ipR}}{4\pi \mathbf{R}} = ip \sum_{l,m} h_l(pr) j_l(pr') Y_l^m(\hat{\mathbf{r}}) Y_l^{m*}(\hat{\mathbf{r}}') , \tag{3.22}$$

the free-particle Green's function becomes [53]:

$$G_0(\mathbf{r}, \mathbf{r}', E) = -ip \sum_{\kappa,\mu} \left[J_\kappa^\mu(\mathbf{r}) H_\kappa^{\mu+\times}(\mathbf{r}')\Theta(r'-r) + H_\kappa^{\mu+}(\mathbf{r}) J_\kappa^{\mu\times}(\mathbf{r}')\Theta(r-r') \right] , \tag{3.23}$$

with

$$J_\kappa^\mu(\mathbf{r}) = \sqrt{\frac{E + mc^2}{c^2}} \left(\begin{array}{c} j_l(pr)\chi_\kappa^\mu(\hat{\mathbf{r}}) \\ \frac{ipcS_\kappa}{E+mc^2} j_{\bar{l}}(pr)\chi_{-\kappa}^\mu(\hat{\mathbf{r}}) \end{array} \right) \tag{3.24}$$

$$H_\kappa^{\mu+}(\mathbf{r}) = \sqrt{\frac{E + mc^2}{c^2}} \left(\begin{array}{c} h_l^+(pr)\chi_\kappa^\mu(\hat{\mathbf{r}}) \\ \frac{ipcS_\kappa}{E+mc^2} h_{\bar{l}}^+(pr)\chi_{-\kappa}^\mu(\hat{\mathbf{r}}) \end{array} \right) . \tag{3.25}$$

In Eq. (3.23) the symbol "$\times$" indicates the left-hand side solutions of the Dirac equation [54]. In the case of free particles, non-magnetic muffin-tin

15

potentials and muffin-tin potentials with magnetization parallel the z-axis the symbol "$\times$" implies the relation:

$$A_\kappa^{\mu\times}(\mathbf{r}) = \begin{pmatrix} g_\kappa(pr)\chi_\kappa^\mu(\hat{\mathbf{r}}) \\ if_\kappa(pr)\chi_{-\kappa}^\mu(\hat{\mathbf{r}}) \end{pmatrix}^\times = \left(g_\kappa(pr)\chi_\kappa^{\mu\dagger}(\hat{\mathbf{r}}), -if_\kappa(pr)\chi_{-\kappa}^{\mu\dagger}(\hat{\mathbf{r}}) \right) . \quad (3.26)$$

As one can see from Eqs. (3.24) and (3.25) the radial dependency of the free-particle Green's function is fully determined through the solutions of the spherical Bessel differential equation and the angular dependency is determined through the spin-angular functions.

3.1.3 Single-Site Scattering

In order to describe a scattering experiment that consists of a free particle which gets scattered by a single potential well $V(\mathbf{r})$ one can use the Lippmann-Schwinger equation [55]:

$$\psi(\mathbf{r}) = \phi(\mathbf{r}) + \int d^3r' G_0(\mathbf{r}, \mathbf{r}')V(\mathbf{r}')\psi(\mathbf{r}') , \quad (3.27)$$

with $\phi(\mathbf{r})$ the solution of the Dirac equation for the free-particle (energy dependencies are omitted). The physical interpretation of this equation is that $\phi(\mathbf{r})$ represents the incident particle and the second term corresponds to an implicit expression for the perturbation of the wave function due to the potential.

If one expands Eq. (3.27):

$$\psi(\mathbf{r}) = \phi(\mathbf{r}) + \int d^3r' G_0(\mathbf{r}, \mathbf{r}')V(\mathbf{r}')\phi(\mathbf{r}') \quad (3.28)$$

$$+ \int\int d^3r' d^3r'' G_0(\mathbf{r}, \mathbf{r}')V(\mathbf{r}')G_0(\mathbf{r}', \mathbf{r}'')V(\mathbf{r}'')\phi(\mathbf{r}'') + ... , \quad (3.29)$$

one can reformulate the Lippmann-Schwinger equation in the following way:

$$\psi(\mathbf{r}) = \phi(\mathbf{r}) + \int\int d^3r' d^3r'' G_0(\mathbf{r}, \mathbf{r}')t(\mathbf{r}', \mathbf{r}'')\phi(\mathbf{r}'') , \quad (3.30)$$

with the t-matrix:

$$t(\mathbf{r}, \mathbf{r}') = V(\mathbf{r})\delta(\mathbf{r} - \mathbf{r}') + V(\mathbf{r})G_0(\mathbf{r}, \mathbf{r}')V(\mathbf{r}') + ... , \quad (3.31)$$

or alternatively:

$$\psi(\mathbf{r}) = \phi(\mathbf{r}) + \int d^3r' G(\mathbf{r}, \mathbf{r}')V(\mathbf{r}')\phi(\mathbf{r}') , \quad (3.32)$$

with

$$G(\mathbf{r}, \mathbf{r}') = G_0(\mathbf{r}, \mathbf{r}') + \int d^3 r'' G_0(\mathbf{r}, \mathbf{r}'') V(\mathbf{r}'') G_0(\mathbf{r}'', \mathbf{r}') + \dots \tag{3.33}$$

$$= G_0(\mathbf{r}, \mathbf{r}') + \int d^3 r'' G_0(\mathbf{r}, \mathbf{r}'') V(\mathbf{r}'') G(\mathbf{r}'', \mathbf{r}') . \tag{3.34}$$

Eq. (3.34) is the so-called Dyson equation.

Using the Dirac notation the Lippmann-Schwinger equation can be written as:

$$|\psi\rangle = |\phi\rangle + G_0 V |\psi\rangle \tag{3.35}$$

$$= \sum_\gamma a_\gamma |\phi_\gamma\rangle + \sum_\gamma G_0 |\phi_\gamma\rangle \langle \phi_\gamma | V |\psi\rangle \tag{3.36}$$

$$= \sum_\gamma a_\gamma |\phi_\gamma\rangle + \sum_\gamma G_0 \, t_\gamma |\phi_\gamma\rangle , \tag{3.37}$$

with the matrix element $t_\gamma = \langle \phi_\gamma | V |\psi\rangle$ or in the $|\mathbf{r}\rangle$ representation:

$$\langle \mathbf{r} | \psi \rangle = \langle \mathbf{r} | \phi \rangle + \int \int d^3 r' d^3 r'' \langle \mathbf{r} | G_0 | \mathbf{r}' \rangle \langle \mathbf{r}' | V | \mathbf{r}'' \rangle \langle \mathbf{r}'' | \psi \rangle \tag{3.38}$$

$$= \langle \mathbf{r} | \phi \rangle$$

$$+ \sum_\gamma \int \int \int d^3 r' d^3 r'' d^3 r''' \langle \mathbf{r} | G_0 | \mathbf{r}' \rangle \langle \mathbf{r}' | \phi_\gamma \rangle \langle \phi_\gamma | \mathbf{r}''' \rangle \langle \mathbf{r}''' | V | \mathbf{r}'' \rangle \langle \mathbf{r}'' | \psi \rangle$$

$$\tag{3.39}$$

$$= \langle \mathbf{r} | \phi \rangle + \sum_\gamma \int d^3 r' \langle \mathbf{r} | G_0 | \mathbf{r}' \rangle t_\gamma \langle \mathbf{r}' | \phi_\gamma \rangle , \tag{3.40}$$

with

$$t_\gamma = \int \int d^3 r d^3 r' \langle \phi_\gamma | \mathbf{r} \rangle \langle \mathbf{r} | V | \mathbf{r}' \rangle \langle \mathbf{r}' | \psi \rangle \delta(\mathbf{r} - \mathbf{r}') . \tag{3.41}$$

In Eq. (3.41) it has been used that the potential is diagonal in $\mathbf{r}$ and $\mathbf{r}'$.

In order to calculate the scattering solution $\psi(\mathbf{r})$ it is difficult to use the Lippmann-Schwinger equation due to the fact that one can use this equation only within an iterative procedure. However, it is possible to calculate $\psi(\mathbf{r})$ in a different way which avoids an iterative expansion of $\psi(\mathbf{r})$ in terms of $\phi(\mathbf{r})$. The idea of this procedure is based on a division of space into two different regions. If the scattering potential $V(\mathbf{r})$ consists of a muffin tin potential with a muffin tin radius r_{mt} the potential $V(\mathbf{r}) \neq 0$ if $r < r_{mt}$ and becomes 0 if $r \geq r_{mt}$. Therefore, the wave function for $r \geq r_{mt}$ must be a linear combination of the regular and irregular zero-potential solutions like

17

Eq. (3.14). In appendix A it is shown that there is a close connection between the coefficients $\cos \delta_\kappa$, $\sin \delta_\kappa$ and t_γ. This decomposition of space allows to solve the Dirac equation in each spatial region independently and to match afterwards the solutions at the boundary (r_{mt}) to obtain a solution which is valid for the whole space. The condition to obtain a smooth matching of the wave functions leads for the phase shift δ_κ to the following expression [52]:

$$\tan \delta_\kappa(E) = \frac{\frac{f_\kappa(r_{\mathrm{mt}},E)}{g_\kappa(r_{\mathrm{mt}},E)} j_l(pr_{\mathrm{mt}}) - \frac{ipcS_\kappa}{E+mc^2} j_{\bar{l}}(pr_{\mathrm{mt}})}{\frac{f_\kappa(r_{\mathrm{mt}},E)}{g_\kappa(r_{\mathrm{mt}},E)} n_l(pr_{\mathrm{mt}}) - \frac{ipcS_\kappa}{E+mc^2} n_{\bar{l}}(pr_{\mathrm{mt}})} . \tag{3.42}$$

The last equation is only applicable if the potential contains no magnetic field. The matching procedure of the wave functions in the magnetic case is similar to the procedure shown above but it becomes more difficult to define appropriated phase shifts (the details are discussed in Ref. [48]).

To evaluate Eq. (3.42) one has to calculate the right-hand side solutions of the Dirac equation for $r < r_{\mathrm{mt}}$. In the following treatment of the Dirac equation the general case of a potential with an additional magnetic field (pointing along the z-axis) is discussed. Using the ansatz from Eq. (3.9) (now with (κ, μ)-dependent radial solutions) in combination with the Dirac equation shown in Eq. (2.17) one obtains [56]:

$$\frac{\partial}{\partial r} P_\Lambda = -\frac{\kappa}{r} P_\Lambda + \left(\frac{E}{c^2} + 1\right) Q_\Lambda - \frac{1}{c^2} \sum_{\Lambda'} V^-_{\Lambda\Lambda'} Q_{\Lambda'} \tag{3.43}$$

$$\frac{\partial}{\partial r} Q_\Lambda = \frac{\kappa}{r} Q_\Lambda - E P_\Lambda + \sum_{\Lambda'} V^+_{\Lambda\Lambda'} P_{\Lambda'} , \tag{3.44}$$

with $P_\Lambda(r, E) = r g_\Lambda(r, E)$, $Q_\Lambda(r, E) = cr f_\Lambda(r, E)$ and the spin-angular matrix elements:

$$V^\pm_{\Lambda\Lambda'}(r) = \langle \chi_{\pm\Lambda} | v_{\mathrm{eff}}(r) \pm \sigma_z B_{\mathrm{eff}}(r) | \chi_{\pm\Lambda'} \rangle \tag{3.45}$$

$(-\Lambda = (-\kappa, \mu))$. If one evaluates the matrix elements given in Eq. (3.45) it turns out that only under certain conditions the matrix elements are $\neq 0$:

$$\langle \chi_\Lambda | \sigma_z | \chi_{\Lambda'} \rangle = \delta_{\mu\mu'} \begin{cases} -\dfrac{\mu}{\kappa + \frac{1}{2}} & \text{for } \kappa = \kappa' \\[2ex] -\sqrt{1 - \dfrac{\mu^2}{\left(\kappa + \frac{1}{2}\right)^2}} & \text{for } \kappa = -\kappa' - 1 \\[2ex] 0 & \text{otherwise} . \end{cases} \tag{3.46}$$

Due to the fact that the allowed values for the quantum number κ are:

$$\kappa = -l - 1 \qquad \text{if} \quad j = l + \frac{1}{2} \tag{3.47}$$

$$\kappa = l \qquad \text{if} \quad j = l - \frac{1}{2} , \tag{3.48}$$

one obtains only a coupling between the solutions of Eqs. (3.43) and (3.44) if $\Delta l = 0, 2$ in combination with $\Delta \mu = 0$. However, this leads to an infinite system of coupled radial differential equations. Feder et al. [47] pointed out that the matrix elements which belong to the coupling $\Delta l = 2$ give a small contribution compared to the matrix elements which belong to the coupling $\Delta l = 0$. Therefore, the coupling $\Delta l = 2$ is neglected in the present work. The coupling $\Delta l = 0$ leads to two coupled radial equations if $j = l + 1/2$ together with $\mu = \pm j$, otherwise one obtains four coupled radial equations. Jenkins and Strange [57] investigated in detail the impact of neglecting the coupling $\Delta l = 2$. They suggested that it may be important to include this coupling if one calculates quantities like the magnetocrystalline anisotropy which are very small on the scale of electronic energies.

3.1.4 The Relativistic Single-Site Scattering Green's Function

To construct the relativistic single-site scattering Green's function one can use the Dyson equation (Eq. (3.34)) in combination with Eqs. (3.31) and (3.23). This leads to the following expression for the Green's function [53]:

$$
G_{ss}(\mathbf{r}, \mathbf{r}', E) = \sum_{\Lambda \Lambda'} Z_\Lambda(\mathbf{r}, E) t_{\Lambda \Lambda'}(E) Z_{\Lambda'}^\times(\mathbf{r}', E)
$$
$$
- \sum_\Lambda \left[Z_\Lambda(\mathbf{r}, E) J_\Lambda^\times(\mathbf{r}', E) \Theta(r' - r) \right.
$$
$$
\left. + J_\Lambda(\mathbf{r}, E) Z_\Lambda^\times(\mathbf{r}', E) \Theta(r - r') \right] , \qquad (3.49)
$$

with

$$
t_{\Lambda \Lambda'}(E) = \int d^3 r \int d^3 r' \, J_\Lambda^\times(\mathbf{r}, E) t(\mathbf{r}, \mathbf{r}', E) J_{\Lambda'}(\mathbf{r}', E) . \qquad (3.50)
$$

For $r, r' \geq r_{\mathrm{mt}}$ one has:

$$
Z_\Lambda(\mathbf{r}, E) = \sum_{\Lambda'} J_{\Lambda'}(\mathbf{r}, E) t_{\Lambda' \Lambda}^{-1}(E) - ip \, H_\Lambda^+(\mathbf{r}, E) \qquad (3.51)
$$

$$
Z_\Lambda^\times(\mathbf{r}', E) = \sum_{\Lambda'} J_{\Lambda'}^\times(\mathbf{r}', E) t_{\Lambda' \Lambda}^{-1}(E) - ip \, H_\Lambda^{+ \times}(\mathbf{r}', E) , \qquad (3.52)
$$

which indicates that Z_Λ and $Z_\Lambda^\times$ are general solutions of the free-particle Dirac equation like Eq. (3.14) but with a different normalization [41, 52]. In the case of $r, r' = r_{\mathrm{mt}}$ these functions match smoothly to the regular right- and left-hand side solutions of the Dirac equation with $V(\mathbf{r}) \neq 0$. As mentioned above, due to this matching procedure at r_{mt} one obtains solutions of the Dirac equation for the whole space.

3.2   Multiple Scattering Theory

In order to describe the propagation of an incident wave through a complex scattering assembly one can use the Lippmann-Schwinger equation as in the case of a single scattering potential:

$$|\psi^{\text{in},i}\rangle = |\phi\rangle + G_0 \sum_{k \neq i} t^k |\psi^{\text{in},k}\rangle \ . \tag{3.53}$$

This equation describes the incoming wave at site i which consists of a superposition of an unperturbed wave $|\phi\rangle$ with incoming waves at site k are scattered there and propagate via G_0 toward site i. The wave function of the entire system can be written as:

$$|\psi\rangle = |\phi\rangle + G_0 T |\phi\rangle \ , \tag{3.54}$$

with the total T-matrix:

$$T = \sum_i t^i + \sum_i \sum_{k \neq i} t^i G_0 t^k + \sum_i \sum_{k \neq i} \sum_{j \neq k} t^i G_0 t^k G_0 t^j + ... \ . \tag{3.55}$$

The exclusions in the sums of Eq. (3.55) prevent two successive scattering events occurring at the same site. This restriction is necessary because the t-matrix describes the complete scattering of a single scattering potential i.e. within an atomic potential well.

An alternative way to write the multiple scattering series shown in Eq. (3.55) is to use the so-called scattering-path operator τ^{ij} which was first introduced by Györffy and Stott [58] and is one of the central quantities of multiple scattering theory. The scattering-path operator consists of a partial summation which describes the complete scattering between the sites i and j. With the help of the scattering-path operator the following expression can be derived for the T-matrix:

$$T = \sum_{i,j} \tau^{ij} \ , \tag{3.56}$$

with

$$\tau^{ij} = t^i \delta_{ij} + \sum_{k \neq i} t^i G_0 t^k \delta_{kj} + \sum_{k \neq i} \sum_{l \neq k} t^i G_0 t^k G_0 t^l \delta_{lj} + ... \ . \tag{3.57}$$

Eq. (3.57) clearly shows the physical interpretation of the scattering-path operator. The operator τ^{ij} describes a scattering event that starts at site j, propagates via G_0 to other scattering sites and finally ends at site i.

In the introduction to this chapter it was mentioned that the decoupling of structural aspects from potential aspects is an advantage of the KKR-GF method. In order to point out this property it is useful to introduce in Eq. (3.57) an expansion of the free-particle Green's function $G_0(\mathbf{r}, \mathbf{r}', E)$ in terms of spherical Bessel functions centered around two different lattice sites $\mathbf{R}_i$, $\mathbf{R}_j$ [52]:

$$
\begin{aligned}
G_0(\mathbf{r}, \mathbf{r}', E) &= G_0(\mathbf{r}_i + \mathbf{R}_i, \mathbf{r}_j + \mathbf{R}_j, E) \\
&= \sum_{\Lambda\Lambda'} J_\Lambda(\mathbf{r}_i, E) G_{0,\Lambda\Lambda'}(\mathbf{R}_i - \mathbf{R}_j, E) J_{\Lambda'}^\times(\mathbf{r}_j, E) ,
\end{aligned}
\tag{3.58}
$$

where the expansion coefficients $G_{0,\Lambda\Lambda'}(\mathbf{R}_i - \mathbf{R}_j, E)$ with $\mathbf{R}_i \neq \mathbf{R}_j$ are the so-called real-space structure constants. The real-space structure constants consist of Gaunt coefficients in combination with Hankel functions (explicit expressions are given in Refs. [41, 52]). Inserting Eq. (3.58) into Eq. (3.57) and multiplying from left with $J_{\Lambda'}^\times(\mathbf{r}_i, E)$ and from right with $J_{\Lambda'}(\mathbf{r}_j, E)$ in combination with an integration over $\mathbf{r}_i$ and $\mathbf{r}_j$ one obtains:

$$
\tau_{\Lambda\Lambda'}^{ij}(E) = t_{\Lambda\Lambda'}^i(E)\delta_{ij} + \sum_{k \neq i} \sum_{\Lambda'\Lambda''} t_{\Lambda\Lambda'}^i(E) G_{0,\Lambda'\Lambda''}(\mathbf{R}_i - \mathbf{R}_k, E)\tau_{\Lambda''\Lambda'}^{kj}(E) ,
\tag{3.59}
$$

or in matrix notation:

$$
\underline{\underline{\tau}} = \underline{\underline{t}} + \underline{\underline{t}}\,\underline{\underline{G}}_0\,\underline{\underline{\tau}} ,
\tag{3.60}
$$

where $_=$ indicates a supermatrix with respect to the site and angular momentum indices. It is important to note that the matrix $\underline{\underline{t}}$ is diagonal in the site indices ($\underline{t}^i \delta_{ij}$) whereas the matrix $\underline{\underline{G}}_0$ consists only of non-diagonal site elements ($\underline{G}_0^{ij}(1 - \delta_{ij})$). Eq. (3.60) can be solved for finite systems via a matrix inversion:

$$
\underline{\underline{\tau}} = \left(\underline{\underline{t}}^{-1} - \underline{\underline{G}}_0\right)^{-1} .
\tag{3.61}
$$

This equation clearly shows the decoupling of the potential aspects which enter into the determination of the t-matrix from the structural aspects which determine the structure constants.

3.2.1 The Relativistic Multiple Scattering Green's Function

For the derivation of the multiple scattering Green's function one can start from the Dyson equation:

$$
G^{nn} = G_{ss}^n + G_{ss}^n T^{nn} G_{ss}^n ,
\tag{3.62}
$$

with

$$T^{nn} = \sum_{i \neq n} \sum_{j \neq n} \tau^{ij} \; . \tag{3.63}$$

The idea of Eq. (3.62) is that one starts with a reference system that consists of a single scatterer at position n surrounded by vacuum. This reference system is fully described by the single-site scattering Green's function G_{ss}^n. The remaining scattering centers are treated as a perturbation. This method for the determination of the multiple scattering Green's function was first suggested by Faulkner and Stocks [59] for the non-relativistic case.

Finally, one obtains for the relativistic multiple scattering Green's function [53]:

$$
\begin{aligned}
G(\mathbf{r}, \mathbf{r}', E) = & \sum_{\Lambda \Lambda'} Z_\Lambda^i(\mathbf{r}, E) \tau_{\Lambda \Lambda'}^{ij}(E) Z_{\Lambda'}^{j \times}(\mathbf{r}', E) \\
& - \sum_{\Lambda} \left[Z_\Lambda^i(\mathbf{r}, E) J_\Lambda^{i \times}(\mathbf{r}', E) \Theta(r' - r) \right. \\
& \left. + J_\Lambda^i(\mathbf{r}, E) Z_\Lambda^{i \times}(\mathbf{r}', E) \Theta(r - r') \right] \delta_{ij} \; ,
\end{aligned}
\tag{3.64}
$$

for $\mathbf{r}$ within the cell i and $\mathbf{r}'$ within the cell j. This expression is very similar to the expression for the single-site scattering Green's function (Eq. (3.49)). The only difference compared to Eq. (3.49) is that the t-matrix is replaced by the scattering-path operator τ.

3.2.2 Coherent Potential Approximation (CPA)

The problem of calculating the electronic structure of a random substitutional alloy is a classic problem in solid state physics [60]. The theoretical description of alloys is hindered due to the fact that traditional methodologies of solid state physics are no longer applicable because of the loss of translational symmetry [61].

The best single-site approach to describe random substitutional alloys is the CPA [41]. The CPA was first introduced by Soven [62] and is a mean field theory which leads to the construction of an effective medium that mimics the scattering properties of an alloy in an averaged way.

The configuration averaged Green's function can be written as [53]:

$$\langle G(E) \rangle = \langle (E - H)^{-1} \rangle = [E - H_0 - \Sigma(E)]^{-1} \; , \tag{3.65}$$

or via a Dyson equation:

$$\langle G(E) \rangle = G_0 + G_0 \Sigma(E) \langle G(E) \rangle \; , \tag{3.66}$$

with the electron self-energy operator $\Sigma(E)$ and $H = H_0 + \sum_i V_i$. For a binary alloy $A_x B_{1-x}$ the potential at site i is either V_A or V_B. If one now introduces an energy dependent translational invariant site quantity $\mathcal{W}_i(E)$ one can rewrite the expression for the Green's function in the following way:

$$G(E) = [E - H_0 - V + \mathcal{W}(E) - \mathcal{W}(E)]^{-1} = [E - \gamma(E) - \mathcal{H}(E)]^{-1} , \quad (3.67)$$

with $\gamma(E) = \sum_i [V_i - \mathcal{W}_i(E)] = V - \mathcal{W}(E)$ and $\mathcal{H}(E) = H_0 + \mathcal{W}(E)$. The consideration of $\gamma(E)$ as a perturbation leads to the "unperturbed" Green's function:

$$\tilde{G}_0(E) = [E - \mathcal{H}(E)]^{-1} . \quad (3.68)$$

With the help of the last equation one can construct a Dyson equation for the full Green's function:

$$G(E) = \tilde{G}_0(E) + \tilde{G}_0(E) T(E) \tilde{G}_0(E) , \quad (3.69)$$

with

$$T(E) = \gamma(E) + \gamma(E) \tilde{G}_0(E) T(E) . \quad (3.70)$$

The averaging over G leads to:

$$\langle G(E) \rangle = \tilde{G}_0(E) + \tilde{G}_0(E) \langle T(E) \rangle \tilde{G}_0(E) , \quad (3.71)$$

because $\tilde{G}_0(E)$ is translationally invariant. It is important to note that Eq. (3.71) is an exact equation. The CPA demands that $\langle T(E) \rangle = 0$ and therefore $\langle G(E) \rangle = \tilde{G}_0(E)$. The alloy is now described via a translationally invariant effective medium $\mathcal{W}(E) = \Sigma(E)$.

The T-matrix can be rewritten with the help of partial summations in a more convenient form:

$$\langle T(E) \rangle = \sum_i \langle \tilde{t}_i(E) \rangle + \left\langle \tilde{t}_i(E) \tilde{G}_0(E) \sum_{j \neq i} Q_j(E) \right\rangle \quad (3.72)$$

$$Q_j(E) = \tilde{t}_j(E) + \tilde{t}_j(E) \tilde{G}_0(E) \sum_{k \neq j} Q_k(E) \quad (3.73)$$

$$\tilde{t}_i(E) = \gamma_i(E) + \gamma_i(E) \tilde{G}_0(E) \tilde{t}_i(E) . \quad (3.74)$$

In the single-site approximation the averaging procedure is restricted to an averaging at single sites i independently of the surrounding sites (e.g. $\langle \tilde{t}_i \tilde{G}_0 t_j \rangle = \langle \tilde{t}_i \rangle \tilde{G}_0 \langle \tilde{t}_j \rangle$). Therefore, the CPA condition $\langle T(E) \rangle = 0$ reduces to $\langle \tilde{t}_i(E) \rangle = 0 \quad \forall i$ which explicitly excludes the incorporation of short-ranged order effects into the averaging procedure. It is important

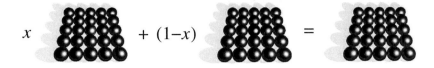

Figure 3.1: The CPA self-consistency condition. The red spheres indicate the CPA medium.

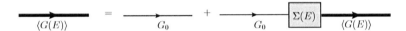

Figure 3.2: Diagrammatic representation of Eq. (3.66).

to note that terms like $\langle \tilde{t}_i \tilde{G}_0 \tilde{t}_j \tilde{G}_0 \tilde{t}_i \tilde{G}_0 \tilde{t}_j \rangle$ in Eq. (3.72) which describe repeated scattering back and forth between a pair of sites are neglected in the CPA because they are not made to vanish by the condition $\langle \tilde{t}_i(E) \rangle = 0 \ \forall i$ ($\langle \tilde{t}_i \tilde{G}_0 \tilde{t}_j \tilde{G}_0 \tilde{t}_i \tilde{G}_0 \tilde{t}_j \rangle \neq \langle \tilde{t}_i \rangle \tilde{G}_0 \langle \tilde{t}_j \rangle \tilde{G}_0 \langle \tilde{t}_i \rangle \tilde{G}_0 \langle \tilde{t}_j \rangle = 0$).

The single-site CPA condition can be also formulated via the scattering-path operator $\langle \tilde{\tau}^{ii}(E) \rangle = 0 \ \forall i$ (Eq. (3.57) with t and G_0 replaced by $\tilde{t}$ and $\tilde{G}_0$, respectively) or alternatively (in the case of a binary alloy $A_x B_{1-x}$) [53]:

$$x \langle \tau^{ii}(E) \rangle_{(i=A)} + (1-x) \langle \tau^{ii}(E) \rangle_{(i=B)} = \tau^{ii}_{\mathrm{CPA}} , \qquad (3.75)$$

where $\langle \tau^{ii}(E) \rangle_{(i=A)}$ is the scattering-path operator of the effective medium with an atom of type A at site i. The physical interpretation of Eq. (3.75) is that excess scattering off a single-site impurity embedded into the effective medium should be zero on the average. A schematic representation of Eq. (3.75) is shown in Fig. 3.1.

The CPA can also be described via diagrammatic techniques. The starting point of a diagrammatic description of the CPA is Eq. (3.66) which can be written in a diagrammatic way as shown in Fig. 3.2. The self-energy $\Sigma(E)$ is defined as the sum of all irreducible diagrams which appear in the expansion of Eq. (3.66) (an irreducible diagram is a diagram which can not be split into two diagrams by cutting a single propagator line G_0) [63]. These diagrams are shown in Fig. 3.3 up to forth order for the case of a statistical independent distribution of the atom types. As mentioned above, the CPA is exact up to the forth order if the atom distribution is purely random. The

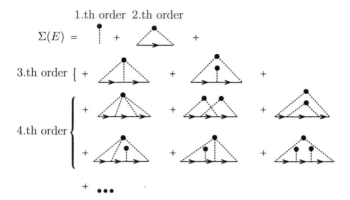

Figure 3.3: Diagrammatic representation of $\Sigma(E)$ up to the forth order. The dashed lines indicate an interaction with a potential at site i which is indicated by the black dot. Different dots represent potentials at different sites. The figure has been taken from Ref. [64].

Figure 3.4: The first diagram neglected in the CPA (all diagrams shown in the present work are created with JaxoDraw [65]).

corresponding neglected diagram is shown in Fig. 3.4 and belongs to the so-called class of crossed diagrams. An appealing feature of the diagrammatic representation is that the differences between the CPA and other alloy theories like the virtual crystal approximation (VCA) or the average t-matrix approximation (ATA) show up in a transparent way (the details of these theories are described for example in Ref. [61]). The VCA self-energy takes only into account the first order diagram shown in Fig. 3.3 whereas the ATA considers also higher order diagrams. The advantage of the CPA compared to the ATA is that also nested diagrams like Fig. 3.5 are included in the CPA.

Figure 3.5: The first nested diagram in the expansion of $\Sigma(E)$.

3.2.3   Non-Local Coherent Potential Approximation (NLCPA)

The major problem of the CPA is that the possibility to investigate correlation effects between the electrostatic potentials on different sites is explicitly excluded by the CPA. Therefore, it is not possible to include e.g. short-ranged ordering effects into the calculations. The self energy expansion shown in Fig. 3.3 is only valid if the atom type distribution is statistical independent. If correlations between the electrostatic potentials on different sites are present diagrams as shown in Fig. 3.6 appear in the self energy expansion. A possible way to overcome the shortcoming of the CPA is to use the

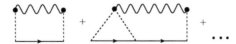

Figure 3.6: Exemplary diagrams for the case of correlated potentials (the correlation is indicated by the wavy line).

non-local coherent potential approximation (NLCPA) [66–69]. The NLCPA has emerged from the dynamical cluster approximation (DCA) [70] which was used originally for the description of dynamical spin and charge fluctuations in strongly correlated electron systems. The static limit of the DCA has been formulated by Jarrell and Krishnamurthy [71] to describe electrons moving in a disordered potential via a tight-binding model.

In order to derive the NLCPA formalism it is useful to start from the CPA. The scattering-path operator within the CPA can be written as [60]:

$$\hat{\underline{\tau}}_{\mathrm{CPA}}^{ij} = \hat{\underline{t}}_{\mathrm{CPA}}\,\delta_{ij} + \sum_{k \neq i} \hat{\underline{t}}_{\mathrm{CPA}}\,\underline{G}(\mathbf{R}^{ik})\,\hat{\underline{\tau}}_{\mathrm{CPA}}^{kj}\,, \qquad (3.76)$$

where the circumflex denotes a quantity of an effective medium and the underscore indicates a matrix in a particular representation (e.g. Λ in the

Figure 3.7: The NLCPA self-consistency condition. The red spheres indicate the NLCPA medium.

relativistic case). In addition, in Eq. (3.76) the structure constants $\underline{G}(\mathbf{R}^{ik})$ appear as function of the position vector $\mathbf{R}^{ij} = \mathbf{R}^i - \mathbf{R}^j$ which connects site i and j.

A similar equation as Eq. (3.76) can be formulated for the NLCPA [66]:

$$\hat{\underline{t}}^{ij}_{\text{NLCPA}} = \hat{\underline{t}}_{\text{NLCPA}}\, \delta_{ij} + \sum_{k \neq i} \hat{\underline{t}}_{\text{NLCPA}} \left[\underline{G}(\mathbf{R}^{ik}) + \delta\hat{\underline{G}}^{ij} \right] \hat{\underline{t}}^{kj}_{\text{NLCPA}} \qquad (3.77)$$

(in the following the acronym NLCPA is omitted). The matrix $\delta\hat{\underline{G}}^{ij}$ is a translationally invariant effective disorder term which takes into account the nonlocal correlations in the investigated material. Due to the fact that the NLCPA is a cluster generalization of the CPA it has to be fulfilled that, similar as in the CPA (Eq. (3.75)), the embedding of an impurity cluster into the NLCPA medium leads to no excess scattering on the average:

$$\sum_{\gamma} P_{\gamma} \underline{\tau}^{IJ}_{\gamma} = \hat{\underline{\tau}}^{IJ}\,, \qquad (3.78)$$

with $\sum_{\gamma} P_{\gamma} = 1$. $\underline{\tau}^{IJ}_{\gamma}$ describes the scattering from cluster site I to cluster site J (capital letters always indicate sites within the cluster) for a cluster with the configuration γ embedded into the NLCPA medium. For a binary alloy the number of possible configurations γ is 2^{N_c} with N_c being the number of sites within the cluster. A schematic picture of the self-consistency condition from Eq. (3.78) is shown in Fig. 3.7 The translational invariance of the NLCPA medium allows it to write the scattering-path operator as [66]:

$$\hat{\underline{\tau}}^{IJ} = \frac{1}{\Omega_{BZ}} \int_{\Omega_{BZ}} d^3k\, [\hat{\underline{t}}^{-1} - \underline{G}(\mathbf{k}) - \delta\hat{\underline{G}}(\mathbf{k})]^{-1}\, e^{i\mathbf{k}(\mathbf{R}^I - \mathbf{R}^J)}\,, \qquad (3.79)$$

with Ω_{BZ} as the volume of the Brillouin zone. Due to the fact that it is not feasible to solve the problem exactly one has to introduce a coarse graining procedure in the spirit of the DCA. The finite number of cluster atoms (N_c) at the sites $\{I\}$ leads to a corresponding set of cluster momenta $\{\mathbf{K}_n\}$ in reciprocal space. The coarse graining amounts to finding $\{I\}$ and $\{\mathbf{K}_n\}$ satisfying:

$$\frac{1}{N_c} \sum_{\mathbf{K}_n} e^{i\mathbf{K}_n(\mathbf{R}^I - \mathbf{R}^J)} = \delta_{IJ} \,. \tag{3.80}$$

This procedure leads to a subdivision of reciprocal space into non-overlapping tiles centered around the vectors $\{\mathbf{K}_n\}$, with N_c tiles covering the Brillouin zone volume [72]. The coarse graining procedure allows to approximate $\delta\hat{\underline{G}}(\mathbf{k})$ through $\delta\hat{\underline{G}}(\mathbf{K}_n)$ if $\mathbf{k}$ is in the nth reciprocal space patch. Finally, the Fourier transform $\delta\hat{\underline{G}}(\mathbf{k})$ and $\delta\hat{\underline{G}}^{IJ}$ are related via the following equations:

$$\delta\hat{\underline{G}}^{IJ} = \frac{1}{N_c} \sum_{\mathbf{K}_n} \delta\hat{\underline{G}}(\mathbf{K}_n) \, e^{i\mathbf{K}_n(\mathbf{R}^I - \mathbf{R}^J)} \tag{3.81}$$

$$\delta\hat{\underline{G}}(\mathbf{K}_n) = \sum_{I \neq J} \delta\hat{\underline{G}}^{IJ} e^{-i\mathbf{K}_n(\mathbf{R}^I - \mathbf{R}^J)} \,. \tag{3.82}$$

An important point to mention is that the NLCPA shows proper behavior for $N_c \to \infty$ and $N_c = 1$, respectively. For N_c going to infinity the NLCPA becomes exact $\{\mathbf{K}_n\} \to \{\mathbf{k}\}$ and correlations over all length scales are included. In the case of $N_c = 1$ the NLCPA recovers the CPA due to the fact that $\delta\hat{\underline{G}}^{II} = 0$.

In order to proceed with Eq. (3.79) one has to apply the coarse graining procedure to the scattering-path operator:

$$\hat{\underline{\tau}}^{IJ} = \frac{1}{\Omega_{BZ}} \sum_{\mathbf{K}_n} \int_{\Omega_{\mathbf{K}_n}} d^3k [\underline{t}^{-1} - \underline{G}(\mathbf{k}) - \delta\hat{\underline{G}}(\mathbf{K}_n)]^{-1} e^{i\mathbf{K}_n(\mathbf{R}^I - \mathbf{R}^J)} \,. \tag{3.83}$$

The NLCPA medium can be iteratively constructed until Eqs. (3.78) and (3.83) are satisfied [66].

A crucial point of the NLCPA is the determination of an appropriate cluster. Jarrell and Krishnamurthy [71] pointed out that the allowed set $\{\mathbf{K}_n\}$ is limited by the point-group symmetry of the Bravais lattice i.e. the real space cluster must preserve the translational symmetry of the lattice. These restrictions lead to the fact that for a fcc lattice the smallest possible cluster consists of four sites ($N_c = 4$) whereas the smallest bcc cluster contains only two sites ($N_c = 2$) [66]. The corresponding number of possible configurations in the case of a binary alloy is 16 (fcc) and 4 (bcc), respectively. In the present work only the smallest fcc and bcc clusters were used due to computational limitations. The details of the cluster construction are discussed in Refs. [66, 67, 73].

3.2.4 Calculating Properties with the Green's Function

The Density of States

In order to calculate the density of states (DOS) one can use the retarded Green's function G^+ (Eq. (3.19)) in the spectral representation and uses the Dirac identity [49]:

$$\lim_{\eta \to 0^+} \int \frac{f(x)}{x \pm i\eta} dx = \left(P \int \frac{f(x)}{x} dx \right) \mp i\pi f(0) , \qquad (3.84)$$

or symbolically

$$\lim_{\eta \to 0^+} \frac{1}{x \pm i\eta} = P\frac{1}{x} \mp i\pi\delta(x) , \qquad (3.85)$$

where P denotes the principle part. This leads to the following expression for the DOS:

$$n(E) = -\frac{1}{\pi}\Im \int \mathrm{Tr}\, G(\mathbf{r}, \mathbf{r}, E)\, d^3r = \sum_n \delta(E - \lambda_n) , \qquad (3.86)$$

where λ_n are the electronic eigenvalues of the system.

The Bloch Spectral Function

In a most general way the density of states may be defined as:

$$n(E) = \sum_n \delta(E - \lambda_n), \qquad (3.87)$$

In analogy the Bloch spectral function (BSF) can be defined by [59]:

$$A(E, \mathbf{k}) = \sum_n \delta[E - \lambda_n(\mathbf{k})] \qquad (3.88)$$

and for that reason can be regarded as a **k**-resolved density of state. Dealing with an ordered system and a given **k**-vector the BSF has at the positions of the eigenvalues an infinitely sharp peak and is zero everywhere else. If one deals with an alloy instead of a perfect crystal an appropriate expression for the BSF within KKR-CPA was worked out by Faulkner and Stocks [59]:

$$
\begin{aligned}
A(E, \mathbf{k}) \;=\; & -\frac{1}{\pi}\Im \mathrm{Tr}\left[\underline{F}^{cc}\underline{\tau}_{\mathrm{CPA}}(E, \mathbf{k})\right] \\
& -\frac{1}{\pi}\Im \mathrm{Tr}\left[(\underline{F}^{c} - \underline{F}^{cc})\underline{\tau}_{\mathrm{CPA}}\right] ,
\end{aligned}
\qquad (3.89)
$$

29

with

$$\underline{\tau}_{\mathrm{CPA}} = \frac{1}{\Omega_{\mathrm{BZ}}} \int_{\mathrm{BZ}} d^3 k \, \underline{\tau}_{\mathrm{CPA}}(E, \mathbf{k}) \, . \tag{3.90}$$

The matrices $\underline{F}^c$ and $\underline{F}^{cc}$ are given in terms of the overlap integrals:

$$F_{\Lambda\Lambda'}^{\alpha\beta} = \int_{\Omega} d^3 r \, Z_{\Lambda}^{\alpha\times}(E, \mathbf{r}) \, Z_{\Lambda'}^{\beta}(E, \mathbf{r}) \, . \tag{3.91}$$

α, β denotes an atom type of the alloy. For more details and explicit expressions see Ref. [59]. Compared to a pure system, the BSF for an alloy becomes broadened due to the disorder. This broadening can be related to the lifetime of an electron in a Bloch state and is therefore quite useful for the interpretation of resistivity data [74].

With the BSF it is possible to discuss a dispersion relation $E(\mathbf{k})$ even for alloys [75]. Strictly spoken such a dispersion relation is in general not defined for alloys because $\mathbf{k}$ is not a good quantum number for disordered systems. Nevertheless, the dispersion relation represented by the BSF can be used to calculate Fermi velocities [74] and gives therefore useful hints for the interpretation of resistivity data.

The derivation of the Bloch spectral function within the NLCPA is presented in Ref. [76].

The Charge Density

The charge density

$$\rho(\mathbf{r}) = \sum_{i=1}^{N} \phi_i^{\dagger}(\mathbf{r}) \phi_i(\mathbf{r}) \tag{3.92}$$

(with N as the number of occupied states) can be easily calculated via the Green's function. If one applies the Dirac identity shown in Eq. (3.85) to the retarded Green's function in the spectral representation one obtains:

$$\rho(\mathbf{r}) = -\frac{1}{\pi} \Im \int_{-\infty}^{E_F} \mathrm{Tr} \, G(\mathbf{r}, \mathbf{r}, E) \, dE \, . \tag{3.93}$$

Chapter 4

Electronic Transport within the Kubo formalism

4.1 Kubo Equation

Many experiments in condensed matter physics investigate the response of a system under an external perturbation. The perturbation can e.g. be an electric or a magnetic field. In the case of a weak perturbation the response of the investigated system is often directly proportional to the intensity of the external perturbation (linear response). In order to investigate such an experiment from a theoretical point of view the Kubo formalism can be employed. The Kubo formalism contains a whole class of equations which describe in a rigorous quantum mechanical way the linear response in terms of correlation functions of the perturbation and the response. Such an equation was first proposed by Green [77] in order to investigate transport in liquids. Kubo [78] was the first who derived such an equation for the calculation of the electrical conductivity in solids.

Due to the fact that the Kubo formalism is based on linear response theory the subsequent procedure is only valid for small perturbations. The starting point for the derivation of the Kubo equation is a system which is in equilibrium and described by the Hamiltonian $\hat{\mathcal{H}}$. On this system a small time dependent perturbation $\hat{W}(t)$ e.g. an electric field (adiabatic switching on) is applied. The question which arises is how an expectation value of an arbitrary operator $\hat{D}$ transforms due to the perturbation? To answer this question one needs the statistical operator $\rho(t)$. If the system is in equilibrium one can calculate the expectation value of an arbitrary operator $\hat{D}$ via the formula [23]:

$$\langle \hat{D} \rangle = \mathrm{Tr}(\rho_0 \hat{D}) , \qquad (4.1)$$

with the density matrix:

$$\rho_0 = \frac{e^{-\beta\hat{\mathcal{H}}}}{\text{Tr}(e^{-\beta\hat{\mathcal{H}}})} \ . \tag{4.2}$$

Here $\beta = (k_B T)^{-1}$, k_B is the Boltzmann constant and T is the temperature. After applying a time dependent perturbation to the system the expectation value of the operator $\hat{D}$ becomes time dependent. In order to calculate the corresponding time dependent expectation value one uses the time dependent density matrix $\rho(t)$:

$$< \hat{D} >_t = \text{Tr}[\rho(t)\hat{D}] \ . \tag{4.3}$$

To find an expression for $\rho(t)$ one can use the fact that $\rho(t)$ fulfills the von-Neumann-equation:

$$i\hbar\frac{\partial}{\partial t}\rho(t) = [(\hat{\mathcal{H}} + \hat{W}(t)), \rho(t)] \ . \tag{4.4}$$

The von-Neumann-equation consists of a commutator between the total Hamiltonian of the perturbed system $\hat{H}$ and $\rho(t)$. In Eq. (4.4) $\hat{H}$ is decomposed into a part which describes the unperturbed system $\hat{\mathcal{H}}$ and a part which describes the perturbation $\hat{W}(t)$. If $\rho(t)$ is transformed from the Schrödinger picture to the interaction picture:

$$\rho_I(t) = e^{i\hat{\mathcal{H}}t/\hbar}\rho(t)e^{-i\hat{\mathcal{H}}t/\hbar} \ , \tag{4.5}$$

one can rewrite the von-Neumann-equation in the following way:

$$i\hbar\frac{\partial}{\partial t}\rho_I(t) = i\hbar\left\{ i/\hbar[\hat{\mathcal{H}}, \rho_I(t)] + e^{i\hat{\mathcal{H}}t/\hbar}\frac{\partial\rho(t)}{\partial t}e^{-i\hat{\mathcal{H}}t/\hbar} \right\} \tag{4.6}$$

$$= -\left\{ [\hat{\mathcal{H}}, \rho_I(t)] - e^{i\hat{\mathcal{H}}t/\hbar}[(\hat{\mathcal{H}} + \hat{W}(t)), \rho(t)]e^{-i\hat{\mathcal{H}}t/\hbar} \right\} \tag{4.7}$$

$$= [\hat{W}_I(t), \rho_I(t)] \ , \tag{4.8}$$

with $\hat{W}_I(t) = e^{i\hat{\mathcal{H}}t/\hbar}\hat{W}(t)e^{-i\hat{\mathcal{H}}t/\hbar}$. The advantage of the change from the Schrödinger picture to the interaction picture is that in Eq. (4.8) the Hamiltonian of the unperturbed system ($\hat{\mathcal{H}}$) disappears. In a next step one can transform back to the Schrödinger picture and integrate over time:

$$\rho(t) = \rho_0 - i/\hbar\int_{-\infty}^{t} dt'\, e^{-i\hat{\mathcal{H}}(t-t')/\hbar}\,[\hat{W}(t'), \rho(t')]\, e^{i\hat{\mathcal{H}}(t-t')/\hbar} \ . \tag{4.9}$$

This equation shows that one needs for the calculation of $\rho(t)$ the density matrix for previous times ($t' < t$). Due to the fact that small perturbations are considered one can solve this equation iteratively:

$$\rho(t) = \rho_0 + \sum_{n=1}^{\infty} (-i/\hbar)^n \int_{-\infty}^{t} dt_1 \int_{-\infty}^{t_1} dt_2 ... \int_{-\infty}^{t_{n-1}} dt_n \, e^{-i\hat{\mathcal{H}}t/\hbar}$$

$$* [\hat{W}_I(t_1), [\hat{W}_I(t_2), [..., [\hat{W}_I(t_n), \rho_0]...]]] \, e^{i\hat{\mathcal{H}}t/\hbar} . \qquad (4.10)$$

For small perturbations it is justified to keep only the first order term:

$$\rho_1(t) = \rho_0 - i/\hbar \int_{-\infty}^{t} dt' e^{-i\hat{\mathcal{H}}t/\hbar} [\hat{W}_I(t'), \rho_0] \, e^{i\hat{\mathcal{H}}t/\hbar} . \qquad (4.11)$$

The combination of the Eqs. (4.3) and (4.11) leads to:

$$< \hat{D} >_t = \mathrm{Tr}(\rho_0 \hat{D}) - i/\hbar \, \mathrm{Tr} \int_{-\infty}^{t} dt' \, e^{-i\hat{\mathcal{H}}t/\hbar} [\hat{W}_I(t'), \rho_0] \, e^{i\hat{\mathcal{H}}t/\hbar} \hat{D} \qquad (4.12)$$

$$= < \hat{D} > -i/\hbar \int_{-\infty}^{\infty} dt' \, \Theta(t - t') < [\hat{D}_I(t), \hat{W}_I(t')] > . \qquad (4.13)$$

Eq. (4.13) is the central equation within linear response theory. With this equation it is possible to calculate the expectation value of an arbitrary operator for a perturbed system only in terms of the commutator of the unperturbed density matrix ρ_0 with the operator representing the perturbation.

For the calculation of the response to an applied electric field $\mathbf{E}_t = \mathbf{E}_0 e^{-i(\omega+i\delta)t}$ (with $\delta \to 0^+$, the term $i\delta$ takes into account that for $t \to -\infty$ the investigated system is unperturbed) one has $\hat{D} = \hat{\mathbf{j}}$ with the current density operator $\hat{\mathbf{j}}$. Due to the fact that the present work investigates also the spin-current density response (spin Hall effect) in the following the operator $\hat{\mathbf{J}}$ is used to represent either the electric current or the spin-current density as a response of the system. The applied electric field couples to the operator of the electric dipole moment:

$$\hat{\mathbf{P}} = \sum_{i=1}^{N} q_i \hat{\mathbf{r}}_i , \qquad (4.14)$$

with the charge q_i and the position operator $\hat{\mathbf{r}}_i$ for the i-th point charge. This gives an interaction operator $\hat{W}_t = -\hat{\mathbf{P}} \cdot \mathbf{E}_t$ which can be inserted into Eq. (4.13) [49]. This leads to:

$$< \hat{J}_\mu >_t = i/\hbar \sum_\nu \int_{-\infty}^{\infty} dt' \, \Theta(t - t') < [\hat{J}_{\mu,I}(t), \hat{P}_{\nu,I}(t')] > E_{0,\nu} e^{-i(\omega+i\delta)t'} .$$

$$(4.15)$$

The first term in Eq. (4.13) drops out in Eq. (4.15) because no current appears if no perturbation is present. The expectation value of the commutator in Eq. (4.15) can be rewritten in a way that one of the operators remains time dependent:

$$
\begin{aligned}
< [\hat{J}_{\mu,I}(t), \hat{P}_{\nu,I}(t')] > &= \mathrm{Tr} \left(\rho_0 \left[e^{i\hat{H}t/\hbar} \hat{J}_\mu e^{-i\hat{H}t/\hbar} e^{i\hat{H}t'/\hbar} \hat{P}_\nu e^{-i\hat{H}t'/\hbar} \right. \right. \\
&\left. \left. - e^{i\hat{H}t'/\hbar} \hat{P}_\nu e^{-i\hat{H}t'/\hbar} e^{i\hat{H}t/\hbar} \hat{J}_\mu e^{-i\hat{H}t/\hbar} \right] \right) \quad (4.16)
\end{aligned}
$$

$$
\begin{aligned}
&= \mathrm{Tr} \left(\rho_0 \left[\hat{J}_\mu e^{i\hat{H}(t'-t)/\hbar} \hat{P}_\nu e^{-i\hat{H}(t'-t)/\hbar} \right. \right. \\
&\left. \left. - e^{i\hat{H}(t'-t)/\hbar} \hat{P}_\nu e^{-i\hat{H}(t'-t)/\hbar} \hat{J}_\mu \right] \right) \quad (4.17)
\end{aligned}
$$

$$
= < [\hat{J}_\mu, \hat{P}_{\nu,I}(t'-t)] > \; . \quad (4.18)
$$

The last equation can be inserted into Eq. (4.15):

$$
\begin{aligned}
< \hat{J}_\mu >_t &= i/\hbar \sum_\nu \int_{-\infty}^\infty dt' \, \Theta(t-t') \\
&\quad < [\hat{J}_\mu, \hat{P}_{\nu,I}(t'-t)] > e^{-i(\omega+i\delta)(t'-t)} E_{0,\nu} e^{-i(\omega+i\delta)t} \quad (4.19) \\
&= i/\hbar \sum_\nu \int_{-\infty}^\infty dt'' \, \Theta(-t'') \\
&\quad < [\hat{J}_\mu, \hat{P}_{\nu,I}(t'')] > e^{-i(\omega+i\delta)t''} E_{t,\nu} \; . \quad (4.20)
\end{aligned}
$$

From Eq. (4.20) one obtains via the relation $J_\mu = \sigma_{\mu\nu} E_\nu$ for the frequency dependent conductivity tensor:

$$
\sigma_{\mu\nu} = i/\hbar \int_{-\infty}^\infty dt \, \Theta(-t) < [\hat{J}_\mu, \hat{P}_{\nu,I}(t)] > e^{-i(\omega+i\delta)t} \; . \quad (4.21)
$$

The expression for the conductivity tensor can be rewritten with the help of the Kubo identity [49]:

$$
[\hat{O}(t), \rho] = -i\hbar\rho \int_0^{(k_B T)^{-1}} d\lambda \, \dot{\hat{O}}(t - i\hbar\lambda) \; , \quad (4.22)
$$

which is valid for any operator $\hat{O}$. Combining this identity with the relation $\dot{\hat{P}} = V \hat{j}$ (V is the volume of the sample) gives a new expression for the conductivity tensor:

$$\sigma_{\mu\nu} = i/\hbar \int_{-\infty}^{\infty} dt \, \Theta(-t) \mathrm{Tr} \left([\hat{P}_{\nu,I}(t), \rho_0] \hat{J}_\mu \right) e^{-i(\omega+i\delta)t} \tag{4.23}$$

$$= \int_0^{(k_B T)^{-1}} d\lambda \int_{-\infty}^0 dt \, \mathrm{Tr} \left(\rho_0 \dot{\hat{P}}_{\nu,I}(t - i\hbar\lambda) \hat{J}_\mu \right) e^{-i(\omega+i\delta)t} \tag{4.24}$$

$$= V \int_0^{(k_B T)^{-1}} d\lambda \int_{-\infty}^0 dt \, \mathrm{Tr} \left(\rho_0 \hat{j}_{\nu,I}(t - i\hbar\lambda) \hat{J}_\mu \right) e^{-i(\omega+i\delta)t} \tag{4.25}$$

$$= V \int_0^{(k_B T)^{-1}} d\lambda \int_0^{\infty} dt \, \mathrm{Tr} \left(\rho_0 \hat{j}_\nu \hat{J}_{I,\mu}(t + i\hbar\lambda) \right) e^{i(\omega+i\delta)t} \,, \tag{4.26}$$

or

$$\sigma_{\mu\nu} = V \int_0^{(k_B T)^{-1}} d\lambda \int_0^{\infty} dt \, < \hat{j}_\nu \hat{J}_{I,\mu}(t + i\hbar\lambda) > e^{i(\omega+i\delta)t} \,. \tag{4.27}$$

This equation is the so-called Kubo equation [78] for the conductivity corresponding to the general response $\hat{J}$. The Kubo equation primarily consists of a current-current correlation function. For the derivation of Eq. (4.26) one uses again the cyclic permutation under the trace in combination with the observation that the correlation function only depends on the time difference [49]. The Kubo equation is in principle an exact equation and not restricted to the single particle picture. The only restriction of this equation is that only small perturbations are allowed to be investigated. Nevertheless, solving the Kubo equation is a very difficult task because it takes into account all many body effects. To make the scheme tractable, further approximations have been devised one of them using the independent electron approximation. This procedure outlined in the next section leads to the Kubo-Středa equation.

4.2 Kubo-Středa Equation

The Kubo-Středa-equation [79] is derived from the Kubo equation using a single particle picture. A similar equation was derived before by Bastin et al. [80]. The starting point is Eq. (4.26):

$$\sigma_{\mu\nu} = V \int_0^{(k_B T)^{-1}} d\lambda \int_0^{\infty} dt \, \mathrm{Tr} \left(\left\langle \rho_0 \hat{j}_\nu \hat{J}_{I,\mu}(t + i\hbar\lambda) \right\rangle_c \right) e^{i(\omega+i\delta)t} \,, \tag{4.28}$$

where $<>_c$ indicates a configurational average (an average over configurations is required for the investigation of alloys). If one neglects the dependence on

ω and uses the independent electron picture one obtains [81]:

$$\sigma_{\mu\nu} = \frac{1}{V} \int_0^{(k_BT)^{-1}} d\lambda \int_0^\infty dt \sum_{n,m} \left\langle e^{-\lambda(E_n-E_m)} f(E_m)[1 - f(E_n)] \right.$$

$$\left. e^{it/\hbar(i\hbar\delta+E_n-E_m)} < m|\hat{j}_\nu|n><n|\hat{J}_\mu|m > \right\rangle_c , \qquad (4.29)$$

with the Fermi-Dirac distribution function $f(E) = (e^{(E-\mu)/k_BT} + 1)^{-1}$ where μ denotes the chemical potential. Performing the integrations over λ and t leads to:

$$\sigma_{\mu\nu} = \frac{i\hbar}{V} \sum_{n,m} \left\langle \frac{f(E_m) - f(E_n)}{(E_n - E_m)(E_n - E_m + i\hbar\delta)} < m|\hat{j}_\nu|n><n|\hat{J}_\mu|m > \right\rangle_c , \qquad (4.30)$$

where for the integration over t the identity [49]:

$$\Theta(t) = \lim_{\delta\to 0^+} \frac{i}{2\pi} \int_{-\infty}^\infty dx \frac{e^{-ixt}}{x + i\delta} , \qquad (4.31)$$

has been applied. Eq. (4.30) is similar to expressions frequently used for the calculation of the Hall and the spin Hall effect for pure systems [82–84]. These works calculate $\sigma_{\mu\nu}$ via Berry-curvatures (see chapter 7) that arises naturally within the Kubo formalism.

One can use the relation $\int_{-\infty}^\infty dE\delta(E - \hat{H}) = 1$ together with the identity:

$$\lim_{\delta\to 0^+} \frac{1}{(E_n - E)(E_n - E + i\delta)} = \lim_{\delta\to 0^+} \frac{d}{dE} \left(\frac{1}{E_n - E + i\delta} \right) \qquad (4.32)$$

and obtains:

$$\sigma_{\mu\nu} = -\frac{i\hbar}{V} \int_{-\infty}^\infty dE f(E)$$

$$\sum_{n,m} \left\langle < m|\hat{j}_\nu|n > \frac{d}{dE} \left(\frac{1}{E - E_n - i\delta} \right) < n|\hat{J}_\mu|m > \delta(E - E_m) \right.$$

$$\left. - < m|\hat{j}_\nu|n > \delta(E - E_n) < n|\hat{J}_\mu|m > \frac{d}{dE} \left(\frac{1}{E - E_m + i\delta} \right) \right\rangle_c , \qquad (4.33)$$

or similar in operator notation:

$$\sigma_{\mu\nu} = \frac{i\hbar}{V} \int_{-\infty}^{\infty} dE f(E) \text{Tr} \left\langle \hat{J}_\mu \frac{dG^+(E)}{dE} \hat{j}_\nu \delta(E - \hat{H}) - \hat{J}_\mu \delta(E - \hat{H}) \hat{j}_\nu \frac{dG^-(E)}{dE} \right\rangle_c$$

$$(4.34)$$

with $G^\pm(E) = (E - \hat{H} \pm i\delta)^{-1}$. Eq. (4.34) is the so-called Bastin formula [80]. This equation primarily consists of an energy integration over terms which involve the energy derivative of the retarded and advanced Green's functions. The numerical difficulties which appear if one tries to deal with this equation are that one has to calculate the energy derivative of the Green's function which includes the regular as well as the irregular solutions of the Schrödinger/Dirac-equation. In addition, one has to calculate an integral over δ-function like terms which naturally leads to a very slow convergence of the numerical integration. A possible way to overcome these difficulties could be a shift of the integration into the complex plane. Středa [79] reformulated Eq. (4.34) for the athermal limit ($T = 0$) in the Schrödinger case. He obtained an expression without any integration and all energy dependent terms have to be evaluated at the Fermi energy E_F:

$$\sigma_{\mu\nu} = \frac{\hbar}{4\pi V} \text{Tr} \left\langle \hat{J}_\mu (G^+ - G^-) \hat{j}_\nu G^- - \hat{J}_\mu G^+ \hat{j}_\nu (G^+ - G^-) \right\rangle_c$$
$$+ \frac{e}{4\pi i V} \text{Tr} \left\langle (G^+ - G^-)(\hat{r}_\mu \hat{J}_\nu - \hat{r}_\nu \hat{J}_\mu) \right\rangle_c .$$

$$(4.35)$$

This results in the so-called Kubo-Středa equation. Crépieux and Bruno [81] showed that the derivation holds also in the Dirac case.

The derivation of the first term in Eq. (4.35) is straightforward but the second term is written in a way which is only valid under certain circumstances. The second term is a reformulation of the following expression [81]:

$$\tilde{\sigma}_{\mu\nu} = \frac{\hbar}{4\pi V} \int_{-\infty}^{\infty} dE f(E)$$
$$\text{Tr} \left\langle \underbrace{\hat{J}_\mu \frac{dG^-}{dE} \hat{j}_\nu G^- - \hat{J}_\mu G^- \hat{j}_\nu \frac{dG^-}{dE}}_{\tilde{\sigma}_{\mu\nu}^{I,-}} + \underbrace{\hat{J}_\mu G^+ \hat{j}_\nu \frac{dG^+}{dE} - \hat{J}_\mu \frac{dG^+}{dE} \hat{j}_\nu G^+}_{\tilde{\sigma}_{\mu\nu}^{I,+}} \right\rangle_c .$$

$$(4.36)$$

Eq. (4.36) is derived from Eq. (4.34) and can be considered as an equation that consists of two similar integrals which contain only retarded or advanced

Green's functions, respectively. Therefore, it is sufficient to investigate the integral:

$$\tilde{\sigma}_{\mu\nu}^{I} = \frac{\hbar}{4\pi V} \int_{-\infty}^{\infty} dE f(E) \text{Tr} \left\langle \hat{J}_{\mu} \frac{dG}{dE} \hat{j}_{\nu} G - \hat{J}_{\mu} G \hat{j}_{\nu} \frac{dG}{dE} \right\rangle_{c}, \qquad (4.37)$$

without the superscripts $+$ or $-$.

Now one can introduce the velocity operator $\hat{\mathbf{v}}$ via the relation $\hat{\mathbf{j}} = -e\hat{\mathbf{v}}$ ($e = |e|$) into the last equation. For the special case that one is interested in a spin-resolved current response one can use for the operator $\hat{\mathbf{J}}$ the expression $\hat{\mathbf{J}} = -e\mathcal{P}\,\hat{\mathbf{v}}$ which includes additionally a spin-projection operator $\mathcal{P}$ (explicit expressions for $\mathcal{P}$ are derived in Sec. 5.2). The combination of these relations with the identities $i\hbar\hat{v}_i = [\hat{r}_i, \hat{H}] = -[\hat{r}_i, G^{-1}]$ and $\frac{dG}{dE} = -G^2$ leads to:

$$\tilde{\sigma}_{\mu\nu}^{I} = \frac{e}{4\pi i V} \int_{-\infty}^{\infty} dE f(E) \text{Tr} \left\langle \mathcal{P}[\hat{r}_{\mu}, G^{-1}] \frac{dG}{dE} \hat{j}_{\nu} G - \hat{J}_{\mu} G[\hat{r}_{\nu}, G^{-1}] \frac{dG}{dE} \right\rangle_{c}$$

$$(4.38)$$

$$= \frac{e}{4\pi i V} \int_{-\infty}^{\infty} dE f(E) \text{Tr} \left\langle -\mathcal{P}\hat{r}_{\mu} G^{-1} G^2 \hat{j}_{\nu} G + \mathcal{P} G^{-1} \hat{r}_{\mu} G^2 \hat{j}_{\nu} G \right.$$

$$\left. + \hat{J}_{\mu} G \hat{r}_{\nu} G^{-1} G^2 - \hat{J}_{\mu} G G^{-1} \hat{r}_{\nu} G^2 \right\rangle_{c} \qquad (4.39)$$

$$= \frac{e}{4\pi i V} \int_{-\infty}^{\infty} dE f(E)$$

$$\text{Tr} \left\langle -\mathcal{P}\hat{r}_{\mu} G \hat{j}_{\nu} G + \mathcal{P} G^{-1} \hat{r}_{\mu} G^2 \hat{j}_{\nu} G + \hat{J}_{\mu} G \hat{r}_{\nu} G - \hat{J}_{\mu} \hat{r}_{\nu} G^2 \right\rangle_{c} \qquad (4.40)$$

$$= \underbrace{\frac{e}{4\pi i V} \int_{-\infty}^{\infty} dE f(E) \text{Tr} \left\langle G^2 \left(\hat{j}_{\nu} G \mathcal{P} G^{-1} \hat{r}_{\mu} - \hat{J}_{\mu} \hat{r}_{\nu} \right) \right\rangle_{c}}_{=\tilde{\sigma}_{\mu\nu}^{Ia}}$$

$$\underbrace{- \frac{e^2}{4\pi \hbar V} \int_{-\infty}^{\infty} dE f(E) \text{Tr} \left\langle -\mathcal{P}\hat{r}_{\mu} G[\hat{r}_{\nu}, G^{-1}] G + \mathcal{P}[\hat{r}_{\mu}, G^{-1}] G \hat{r}_{\nu} G \right\rangle_{c}}_{=\tilde{\sigma}_{\mu\nu}^{Ib}}.$$

$$(4.41)$$

The first term of Eq. (4.41) can be reformulated as follows:

$$\tilde{\sigma}_{\mu\nu}^{Ia} = \frac{e}{4\pi i V} \int_{-\infty}^{\infty} dE f(E) \text{Tr} \left\langle G^2 \left(\hat{j}_\nu G \mathcal{P} (i\hbar \hat{v}_\mu + \hat{r}_\mu G^{-1}) - \hat{J}_\mu \hat{r}_\nu \right) \right\rangle_c \quad (4.42)$$

$$= \frac{e}{4\pi i V} \int_{-\infty}^{\infty} dE f(E) \text{Tr} \left\langle i\hbar G^2 \hat{j}_\nu G \mathcal{P} \hat{v}_\mu + G \hat{j}_\nu G \mathcal{P} \hat{r}_\mu - G^2 \hat{J}_\mu \hat{r}_\nu \right\rangle_c \quad (4.43)$$

$$= \frac{e}{4\pi i V} \int_{-\infty}^{\infty} dE f(E) \text{Tr} \left\langle i\hbar G^2 \frac{e}{i\hbar} [\hat{r}_\nu, G^{-1}] G \mathcal{P} \hat{v}_\mu \right.$$
$$\left. + G \frac{e}{i\hbar} [\hat{r}_\nu, G^{-1}] G \mathcal{P} \hat{r}_\mu - G^2 \hat{J}_\mu \hat{r}_\nu \right\rangle_c \quad (4.44)$$

$$= \frac{e}{4\pi i V} \int_{-\infty}^{\infty} dE f(E) \text{Tr} \left\langle e G^2 \hat{r}_\nu \mathcal{P} \hat{v}_\mu - e G \hat{r}_\nu G \mathcal{P} \hat{v}_\mu \right.$$
$$\left. + \frac{e}{i\hbar} G \hat{r}_\nu \mathcal{P} \hat{r}_\mu - \frac{e}{i\hbar} \hat{r}_\nu G \mathcal{P} \hat{r}_\mu - G^2 \hat{J}_\mu \hat{r}_\nu \right\rangle_c$$
$$(4.45)$$

$$= \underbrace{\frac{e}{4\pi i V} \int_{-\infty}^{\infty} dE f(E) \text{Tr} \left\langle G^2 \left(e \hat{r}_\nu \mathcal{P} \hat{v}_\mu - \hat{J}_\mu \hat{r}_\nu \right) \right\rangle_c}_{\tilde{\sigma}_{\mu\nu}^{Ia\,G^2}}$$
$$- \frac{e^2}{4\pi i V} \int_{-\infty}^{\infty} dE f(E) \text{Tr} \left\langle G \hat{r}_\nu G \mathcal{P} \hat{v}_\mu \right\rangle_c . \quad (4.46)$$

Under the assumption that the commutation relation $[\mathcal{P}, \hat{\mathbf{r}}] = 0$ concerning the spin-projection operator $\mathcal{P}$ is fulfilled one can reformulate $\tilde{\sigma}_{\mu\nu}^{Ia}$ via a partial integration which gives:

$$\tilde{\sigma}_{\mu\nu}^{Ia\,G^2} = -\frac{e}{2\pi i V} \int_{-\infty}^{\infty} dE f(E) \text{Tr} \left\langle G^2 \hat{J}_\mu \hat{r}_\nu \right\rangle_c \quad (4.47)$$

$$= -\frac{e}{2\pi i V} \int_{-\infty}^{\infty} dE \frac{df(E)}{dE} \text{Tr} \left\langle G \hat{J}_\mu \hat{r}_\nu \right\rangle_c \quad (4.48)$$

$$= \frac{e}{2\pi i V} \int_{-\infty}^{\infty} dE \delta(E - E_F) \text{Tr} \left\langle G \hat{J}_\mu \hat{r}_\nu \right\rangle_c \quad (4.49)$$

$$= \frac{e}{2\pi i V} \text{Tr} \left\langle G(E_F) \hat{J}_\mu \hat{r}_\nu \right\rangle_c . \quad (4.50)$$

Eq. (4.50) is restricted to the athermal case because otherwise the derivative of the Fermi-Dirac distribution function does not become a δ-function.

The second term of Eq. (4.41) gives:

$$\tilde{\sigma}_{\mu\nu}^{Ib} = \frac{e^2}{4\pi\hbar V} \int_{-\infty}^{\infty} dE f(E) \text{Tr} \Big\langle -\mathcal{P}\hat{r}_\mu G \hat{r}_\nu G^{-1} G + \mathcal{P}\hat{r}_\mu G G^{-1} \hat{r}_\nu G$$

$$+ \mathcal{P}\hat{r}_\mu G^{-1} G \hat{r}_\nu G - \mathcal{P}G^{-1}\hat{r}_\mu G \hat{r}_\nu G \Big\rangle_c \quad (4.51)$$

$$= \frac{e^2}{4\pi\hbar V} \int_{-\infty}^{\infty} dE f(E) \text{Tr} \Big\langle -\mathcal{P}\hat{r}_\mu G \hat{r}_\nu + \mathcal{P}\hat{r}_\mu \hat{r}_\nu G$$

$$+ \mathcal{P}\hat{r}_\mu \hat{r}_\nu G - \mathcal{P}G^{-1}\hat{r}_\mu G \hat{r}_\nu G \Big\rangle_c \quad (4.52)$$

$$= \frac{e^2}{4\pi\hbar V} \int_{-\infty}^{\infty} dE f(E) \text{Tr} \Big\langle -\mathcal{P}\hat{r}_\mu G \hat{r}_\nu + \mathcal{P}\hat{r}_\mu \hat{r}_\nu G$$

$$+ \mathcal{P}\hat{r}_\mu \hat{r}_\nu G - \mathcal{P}[i\hbar\hat{v}_\mu + \hat{r}_\mu G^{-1}]G\hat{r}_\nu G \Big\rangle_c \quad (4.53)$$

$$= \frac{e^2}{4\pi\hbar V} \int_{-\infty}^{\infty} dE f(E) \text{Tr} \Big\langle -\mathcal{P}\hat{r}_\mu G \hat{r}_\nu + \mathcal{P}\hat{r}_\mu \hat{r}_\nu G$$

$$+ \mathcal{P}\hat{r}_\mu \hat{r}_\nu G - \mathcal{P}\hat{r}_\mu \hat{r}_\nu G \Big\rangle_c$$

$$+ \frac{e^2}{4\pi i V} \int_{-\infty}^{\infty} dE f(E) \text{Tr} \Big\langle \mathcal{P}\hat{v}_\mu G \hat{r}_\nu G \Big\rangle_c . \quad (4.54)$$

The first integral of the last expression becomes zero when the assumed commutation relation $[\mathcal{P}, \hat{\mathbf{r}}] = 0$ is fulfilled.

Finally, one obtains:

$$\tilde{\sigma}_{\mu\nu} = \tilde{\sigma}_{\mu\nu}^{I,+} + \tilde{\sigma}_{\mu\nu}^{I,-} \tag{4.55}$$

$$= -\frac{e}{2\pi i V}\mathrm{Tr}\Big\langle (G^+ - G^-)\hat{J}_\mu\hat{r}_\nu \Big\rangle_c$$

$$+ \frac{e^2}{2\pi i V}\int_{-\infty}^{\infty} dE\, f(E)\mathrm{Tr}\Big\langle \mathcal{P}\hat{v}_\mu G^+\hat{r}_\nu G^+ - \mathcal{P}\hat{v}_\mu G^-\hat{r}_\nu G^- \Big\rangle_c \tag{4.56}$$

$$= -\frac{e}{2\pi i V}\mathrm{Tr}\Big\langle (G^+ - G^-)\hat{r}_\nu\hat{J}_\mu \Big\rangle_c$$

$$- \frac{e}{2\pi i V}\int_{-\infty}^{\infty} dE\, f(E)\mathrm{Tr}\Big\langle \hat{J}_\mu G^+\hat{r}_\nu G^+ - \hat{J}_\mu G^-\hat{r}_\nu G^- \Big\rangle_c \tag{4.57}$$

$$= \frac{e}{4\pi i V}\mathrm{Tr}\Big\langle (G^+ - G^-)(\hat{r}_\mu\hat{J}_\nu - \hat{r}_\nu\hat{J}_\mu) \Big\rangle_c$$

$$- \frac{e}{2\pi i V}\int_{-\infty}^{\infty} dE\, f(E)\mathrm{Tr}\Big\langle \hat{J}_\mu G^+\hat{r}_\nu G^+ - \hat{J}_\mu G^-\hat{r}_\nu G^- \Big\rangle_c, \tag{4.58}$$

which corresponds to the second term in Eq. (4.35) in combination with an additional Fermi sea term. It is important to note that the Fermi sea term is only present if $[\mathcal{P}, G] \neq 0$.

To summarize, in this section it has been shown how the Kubo-Středa equation can be derived from the Kubo equation within an independent electron picture without any further approximations. The important reformulation of the Bastin formula by Středa [79] avoids an integration over energy which makes the calculation of the Kubo-Středa equation numerically less demanding. All energy dependent terms have to be evaluated at the Fermi energy if $[\mathcal{P}, G] = 0$. In that case only electronic states at the Fermi edge contribute to the conductivity for the athermal limit.

4.3 Kubo-Greenwood Equation

The conductivity tensor of a cubic crystal with spontaneous magnetization along the z-axis has the form [85]:

$$\boldsymbol{\sigma} = \begin{pmatrix} \sigma_{xx} & \sigma_H & 0 \\ -\sigma_H & \sigma_{yy} & 0 \\ 0 & 0 & \sigma_{zz} \end{pmatrix}, \tag{4.59}$$

with the Hall conductivity σ_H and $\sigma_{xx} = \sigma_{yy}$. The diagonal elements of $\boldsymbol{\sigma}$ belong to the symmetric part (for systems with lower symmetries also non-diagonal elements may belong to the symmetric part) and the Hall conductivity belongs to the anti-symmetric part of the conductivity tensor. The Kubo-Středa equation gives access to the complete conductivity tensor whereas the Kubo-Greenwood equation [86] gives only access to the symmetric part. Therefore, it is not possible to calculate the Hall conductivity within the Kubo-Greenwood formalism. For the derivation of the Kubo-Greenwood equation one can start with Eq. (4.35):

$$
\begin{aligned}
\sigma_{\mu\nu} &= \frac{i\hbar}{2\pi V}\mathrm{Tr}\Big\langle \hat{J}_{\mu}\Im G^{+}\hat{j}_{\nu}(\Re G^{+}-i\Im G^{+})-\hat{J}_{\mu}(\Re G^{+}+i\Im G^{+})\hat{j}_{\nu}\Im G^{+}\Big\rangle_{c}\\
&\quad +\frac{e}{2\pi V}\mathrm{Tr}\Big\langle \Im G^{+}(\hat{r}_{\mu}\hat{J}_{\nu}-\hat{r}_{\nu}\hat{J}_{\mu})\Big\rangle_{c} \qquad\qquad (4.60)\\
&= \frac{i\hbar}{2\pi V}\mathrm{Tr}\Big\langle \underbrace{\Big[\hat{J}_{\mu}\Im G^{+}\hat{j}_{\nu}-\hat{j}_{\nu}\Im G^{+}\hat{J}_{\mu}\Big]\Re G^{+}}_{\sigma_{\mu\nu}^{A}}\Big\rangle_{c}\\
&\quad +\underbrace{\frac{\hbar}{\pi V}\mathrm{Tr}\Big\langle \hat{J}_{\mu}\Im G^{+}\hat{j}_{\nu}\Im G^{+}\Big\rangle_{c}}_{\sigma_{\mu\nu}^{B}}\\
&\quad +\underbrace{\frac{e}{2\pi V}\mathrm{Tr}\Big\langle \Im G^{+}(\hat{r}_{\mu}\hat{J}_{\nu}-\hat{r}_{\nu}\hat{J}_{\mu})\Big\rangle_{c}}_{\sigma_{\mu\nu}^{C}} , \qquad\qquad (4.61)
\end{aligned}
$$

with $(G^{+}-G^{-})=2i\Im G^{+}$, $G^{+}=\Re G^{+}+i\Im G^{+}$ and $G^{-}=\Re G^{+}-i\Im G^{+}$. Eq. (4.61) shows that the Kubo-Středa equation consists of three different terms $\sigma_{\mu\nu}=\sigma_{\mu\nu}^{A}+\sigma_{\mu\nu}^{B}+\sigma_{\mu\nu}^{C}$. These terms fulfill the relations $\sigma_{\mu\nu}^{A}=-\sigma_{\nu\mu}^{A}$, $\sigma_{\mu\nu}^{B}=\sigma_{\nu\mu}^{B}$ and $\sigma_{\mu\nu}^{C}=-\sigma_{\nu\mu}^{C}$. This clearly shows that only term $\sigma_{\mu\nu}^{B}$ contributes to the symmetric part of the conductivity tensor. The term $\sigma_{\mu\nu}^{B}$ represents the so-called Kubo-Greenwood equation. This analysis demonstrates that under the assumption of a small Hall conductivity (compared to the symmetric elements) the Kubo-Greenwood equation is sufficient for calculations of the residual resistivity of an alloy.

4.4 Hierarchy of the Linear Response Equations

The following diagram summarizes the various transport equations and their interrelationship:

$$\boxed{\text{Kubo equation}}$$

$$\sigma_{\mu\nu} = V \int_0^{(k_B T)^{-1}} d\lambda \int_0^\infty dt \left\langle \hat{j}_\nu \hat{J}_{I,\mu}(t+i\hbar\lambda) \right\rangle_c e^{i(\omega+i\delta)t}$$

independent electron approximation, $\omega = 0$

$$\boxed{\text{Bastin equation}}$$

$$\sigma_{\mu\nu} = \frac{i\hbar}{V} \int_{-\infty}^\infty dE f(E) \text{Tr} \left\langle \hat{J}_\mu \frac{dG^+(E)}{dE} \hat{j}_\nu \delta(E-\hat{H}) - \hat{J}_\mu \delta(E-\hat{H}) \hat{j}_\nu \frac{dG^-(E)}{dE} \right\rangle_c$$

$T = 0K$

$$\boxed{\text{Kubo-Středa equation}}$$

$$\begin{aligned} \sigma_{\mu\nu} = \quad & \frac{\hbar}{4\pi V} \text{Tr} \left\langle \hat{J}_\mu (G^+ - G^-) \hat{j}_\nu G^- - \hat{J}_\mu G^+ \hat{j}_\nu (G^+ - G^-) \right\rangle_c \\ & + \frac{e}{4\pi i V} \text{Tr} \left\langle (G^+ - G^-)(\hat{r}_\mu \hat{J}_\nu - \hat{r}_\nu \hat{J}_\mu) \right\rangle_c \end{aligned}$$

retaining symmetric part only

$$\boxed{\text{Kubo-Greenwood equation}}$$

$$\frac{\hbar}{\pi V} \text{Tr} \left\langle \hat{J}_\mu \Im G^+ \hat{j}_\nu \Im G^+ \right\rangle_c$$

4.5 Diagrammatic Representation

For the discussion of the conductivity via a diagrammatic representation it is useful to start from Eq. (4.35) in the weak-disorder limit. In the weak-disorder limit the second term from Eq. (4.35) is negligible [87] as well as the contributions of the first term which contains two retarded or two advanced Green's functions (the latter is proved in appendix C for the case of negligible vertex corrections). Therefore, the conductivity reduces to:

$$\sigma_{\mu\nu} = \frac{\hbar}{2\pi V} \mathrm{Tr} \langle \hat{j}_\mu \, G^+ \hat{j}_\nu \, G^- \rangle_c \,. \tag{4.62}$$

This equation can be represented via diagrams as shown in Fig. 4.1. The full

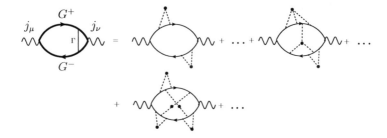

Figure 4.1: The conductivity represented via Feynman diagrams. The meaning of the dashed lines and the black dots is explained in the caption of Fig. 3.3. The curly lines represent the current density operator.

conductivity is illustrated via the diagram which includes the three-point vector vertex Γ [63]. Γ primarily represents all impurity scattering events which connects the two Green's function G^+ and G^- with each other. Therefore, diagrams of the type like the second or the third on the right-hand side in Fig. 4.1 are called vertex diagrams [88] and belong to the class of so-called ladder and crossed diagrams, respectively. Within semiclassical Boltzmann theory the vertex corrections correspond to the scattering-in term (see appendix D) which is important in the case of anisotropic scattering.

In the present work the averaging over two Green's function is carried out in the spirit of the CPA [89] which leads to the neglect of the crossed diagrams which are responsible for the weak-localization effect [63, 90]. If one approximates the averaging over two Green's function in Eq. (4.62) by an averaging over single Green's function:

$$\sigma_{\mu\nu} = \frac{\hbar}{2\pi V} \mathrm{Tr} \, \hat{j}_\mu \langle G^+ \rangle_c \hat{j}_\nu \langle G^- \rangle_c \,, \tag{4.63}$$

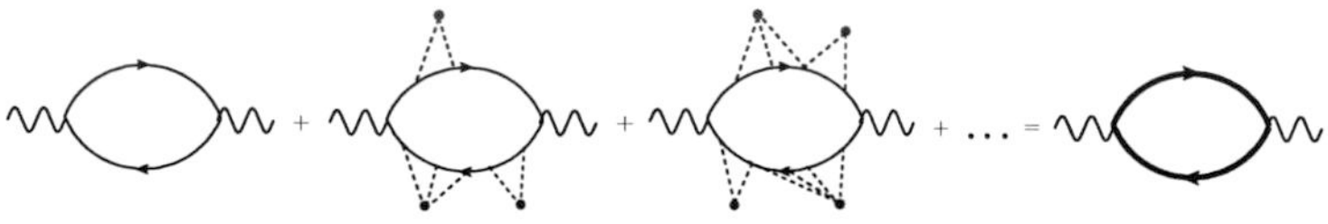

Figure 4.2: Exemplary diagrams contributing to the bubble diagram. The thick lines indicate the averaged propagators $\langle G^{\pm}\rangle_c$.

the vertex diagrams are neglected in the expansion of the full conductivity diagram. Therefore, the conductivity can be represented by the bubble diagram shown in Fig. 4.2. The conductivity calculated via the bubble diagram can be connected with the classical Drude conductivity [63]:

$$\sigma = \frac{e^2 \tau \, n}{m} \, , \tag{4.64}$$

with the electron charge e, the mass m, the carrier density n and the relaxation rate τ.

4.6 Calculation of the conductivity tensor $\sigma_{\mu\nu}$

In this section the details of the calculation of the full conductivity tensor $\sigma_{\mu\nu}$ are presented. The most difficult part in the calculation of the conductivity is the proper inclusion of vertex corrections. The vertex corrections take into account that for the calculation of conductivity one has to perform a configuration average procedure (due to disorder in the investigated alloy) over a product of two Green's function (see Sec. 4.5). Butler [89] derived a scheme to calculate the vertex corrections in a reliable way within the CPA. In this scheme he derives response functions which give access after Fourier transform to the vertex corrections (see Sec. 4.6.1). This non-relativistic scheme was applied by Banhart and Ebert [91] to the spin-polarized relativistic case. Tulip et al. [92] extended Butler's scheme for the inclusion of ordering effects in the lattice site occupation. Therefore, the scheme was reformulated within the NLCPA (see Sec. 4.6.2).

4.6.1 Symmetric Part of $\sigma_{\mu\nu}$ within KKR-CPA

For the calculation of the symmetric part of the conductivity tensor one needs only the Kubo-Greenwood equation:

$$\sigma_{\mu\nu}^{KG} = \frac{\hbar}{\pi N_a \Omega} \operatorname{Tr} \left\langle \hat{J}_\mu \Im G^+(E_F) \hat{j}_\nu \Im G^+(E_F) \right\rangle_c , \qquad (4.65)$$

where N_a is the number of atoms in the system and Ω the volume per atom. Using the identity $\Im G^+ = \frac{1}{2i}(G^+ - G^-)$ the last equation can be decomposed in the following way:

$$\sigma_{\mu\nu}^{KG} = \frac{1}{4} \left[\tilde{\sigma}_{\mu\nu}(G^+, G^+) + \tilde{\sigma}_{\mu\nu}(G^-, G^-) - \tilde{\sigma}_{\mu\nu}(G^+, G^-) - \tilde{\sigma}_{\mu\nu}(G^-, G^+) \right] , \qquad (4.66)$$

with

$$\tilde{\sigma}_{\mu\nu}(G^\pm, G^\pm) = -\frac{\hbar}{\pi N_a \Omega} \operatorname{Tr} \left\langle \hat{J}_\mu G^\pm(E_F) \hat{j}_\mu G^\pm(E_F) \right\rangle_c . \qquad (4.67)$$

For the calculation of $\tilde{\sigma}_{\mu\nu}$ the Green's function shown in Eq. (3.64) has to be inserted into Eq. (4.67). Due to the fact that the second term in Eq. (3.64) is purely real for a real potential and for real energies one can neglect the second term of the standard multiple scattering representation of the Green's function [89]. This leads to:

$$\tilde{\sigma}_{\mu\nu}(z_1, z_2) = -\frac{\hbar}{\pi N_a \Omega} \sum_{m,n} \operatorname{Tr} \left\langle \underline{J}^{m\mu}(z_2, z_1) \underline{\tau}^{mn}(z_1) \underline{j}^{n\mu}(z_1, z_2) \underline{\tau}^{nm}(z_2) \right\rangle_c . \qquad (4.68)$$

Here the underline denotes a matrix in $\Lambda\Lambda'$ with the matrix elements:

$$J_{\Lambda\Lambda'}^{m\mu}(z_2, z_1) = \int d^3r Z_\Lambda^{m\times}(\mathbf{r}, z_2) \, \hat{J}_\mu Z_{\Lambda'}^m(\mathbf{r}, z_1) \qquad (4.69)$$

$$j_{\Lambda\Lambda'}^{n\mu}(z_1, z_2) = \int d^3r Z_\Lambda^{n\times}(\mathbf{r}, z_1) \, \hat{j}_\mu Z_{\Lambda'}^n(\mathbf{r}, z_2) \qquad (4.70)$$

$$= -ec \int d^3r Z_\Lambda^{n\times}(\mathbf{r}, z_1) \, \alpha_\mu Z_{\Lambda'}^n(\mathbf{r}, z_2) , \qquad (4.71)$$

where z_k is a complex number being either $E_F + i\delta$ or $E_F - i\delta$ and $\hat{\mathbf{j}} = -e\hat{\mathbf{v}} = -ec\boldsymbol{\alpha}$ has been used where $\boldsymbol{\alpha}$ is the vector of Dirac matrices (see Eq. (3.2)). The next step in Butler's scheme is to split $\tilde{\sigma}_{\mu\nu}$ into an on-site term centered at site 0:

$$\tilde{\sigma}_{\mu\nu}^0 = -\frac{\hbar}{\pi\Omega} \operatorname{Tr} \left\langle \underline{J}^{0\mu}(z_2, z_1) \underline{\tau}^{00}(z_1) \underline{j}^{0\nu}(z_1, z_2) \underline{\tau}^{00}(z_2) \right\rangle_c \qquad (4.72)$$

and an off-site term:

$$\tilde{\sigma}^1_{\mu\nu} = -\frac{\hbar}{\pi\Omega} \sum_{n\neq 0} \text{Tr}\Big\langle \underline{J}^{0\mu}(z_2,z_1)\underline{\tau}^{0n}(z_1)\underline{j}^{n\nu}(z_1,z_2)\underline{\tau}^{n0}(z_2)\Big\rangle_c . \tag{4.73}$$

The strategy for the further procedure is to define the last two equations via response functions which contain only one current operator and to search for a closed set of auxiliary equations which allows the calculation of these response functions. The necessary response functions K and L are defined via the following equations:

$$\tilde{\sigma}^0_{\mu\nu} = -\frac{\hbar}{\pi\Omega} \sum_{\alpha} x_\alpha \, \text{Tr}\, K^{0\alpha}_\nu(z_1,z_2)\underline{J}^{\alpha\mu}(z_2,z_1) , \tag{4.74}$$

with

$$K^{0\alpha}_\nu(z_1,z_2) = \Big\langle \underline{\tau}^{00}(z_1)\underline{j}^{\alpha\nu}(z_1,z_2)\underline{\tau}^{00}(z_2)\Big\rangle_{0=\alpha} \tag{4.75}$$

and

$$\tilde{\sigma}^1_{\mu\nu} = -\frac{\hbar}{\pi\Omega} \sum_{n\neq 0}\sum_{\alpha,\beta} x_\alpha x_\beta \, \text{Tr}\, L^{0\alpha,n\beta}_\nu(z_1,z_2)\underline{J}^{\alpha\mu}(z_2,z_1) , \tag{4.76}$$

with

$$L^{0\alpha,n\beta}_\nu(z_1,z_2) = \Big\langle \underline{\tau}^{0n}(z_1)\underline{j}^{\beta\nu}(z_1,z_2)\underline{\tau}^{n0}(z_2)\Big\rangle_{0=\alpha,n=\beta} . \tag{4.77}$$

The Eqs. (4.74) - (4.77) contain the indices α and β that give the atom type at a certain lattice position n. In the case of a binary alloy $A_{1-x}B_x$ the matrix element $j^{n\nu}$ must be either that of an A atom (with probability $x_\alpha = 1 - x$) or of a B atom (with probability $x_\beta = x$). Therefore, the notation of the matrix elements is slightly changed i.e. $j^{\beta\nu} = (j^{n\nu})_{n=\beta}$.

Butler was able to show that the response functions are connected within the CPA in the following way:

$$K^{0\alpha}_\nu(z_1,z_2) = \underline{D}^{0\alpha}(z_1)\tilde{K}^{0\alpha}_\nu(z_1,z_2)\underline{\tilde{D}}^{0\alpha}(z_2) , \tag{4.78}$$

with

$$\tilde{K}^{0\alpha}_\nu(z_1,z_2) = \underline{\tau}^{00}_{\text{CPA}}(z_1)\underline{j}^{\alpha\nu}(z_1,z_2)\underline{\tau}^{00}_{\text{CPA}}(z_2)$$
$$+ \sum_{k\neq 0} \underline{\tau}^{0k}_{\text{CPA}}(z_1)\tilde{L}^{k0\alpha}_\nu(z_1,z_2)\,\omega(z_1,z_2)\,\underline{\tau}^{k0}_{\text{CPA}}(z_2) \tag{4.79}$$

and

$$L^{0\alpha,n\beta}_\nu(z_1,z_2) = \underline{D}^{0\alpha}(z_1)\tilde{L}^{0n\beta}_\nu(z_1,z_2)\underline{\tilde{D}}^{0\alpha}(z_2) , \tag{4.80}$$

with

$$\tilde{L}_{\nu}^{0n\beta}(z_1, z_2) = \underline{\tau}_{\text{CPA}}^{0n}(z_1) \underbrace{\underline{D}^{n\beta}(z_1)\underline{j}^{\beta\nu}(z_1, z_2)\underline{\tilde{D}}^{n\beta}(z_2)}_{\tilde{\underline{j}}^{\beta\nu}(z_1,z_2)} \underline{\tau}_{\text{CPA}}^{n0}(z_2)$$

$$+ \sum_{k\neq(0,n)} \underline{\tau}_{\text{CPA}}^{0k}(z_1)\tilde{L}_{\nu}^{kn\beta}(z_1, z_2)\,\omega(z_1, z_2)\,\underline{\tau}_{\text{CPA}}^{k0}(z_1)\,. \qquad (4.81)$$

Here the operators:

$$\underline{D}^{0\alpha}(z) = 1 + \underline{\tau}_{\text{CPA}}^{00}(z)\underline{x}^{\alpha}(z) \qquad (4.82)$$

$$\underline{\tilde{D}}^{0\alpha}(z) = 1 + \underline{x}^{\alpha}(z)\underline{\tau}_{\text{CPA}}^{00}(z) \qquad (4.83)$$

$$\underline{x}^{\alpha}(z) = \left\{ \left[\underline{t}_{\alpha}^{-1}(z) - \underline{t}_{\text{CPA}}^{-1}(z)\right]^{-1} + \underline{\tau}_{\text{CPA}}^{00}(z) \right\} \qquad (4.84)$$

$$\omega_{\Lambda_1\Lambda_2\Lambda_3\Lambda_4}(z_1, z_2) = \sum_{\alpha} x_{\alpha}x_{\Lambda_1\Lambda_2}^{\alpha}(z_1)x_{\Lambda_3\Lambda_4}^{\alpha}(z_2)\,, \qquad (4.85)$$

have been used with $\underline{t}_{\alpha}$ and $\underline{t}_{\text{CPA}}$ are the single site t-matrix for atom type α and for the CPA medium, respectively. It turns out that the only response function which has to be calculated is $\tilde{L}$. Butler [89] shows that this can be done with a Fourier transform of $\tilde{L}$. This procedure leads to an explicit expression for the response function $\tilde{L}$:

$$\tilde{L} = (1 - \chi\omega)^{-1}\chi\tilde{j}\,, \qquad (4.86)$$

with

$$\chi_{\Lambda_1\Lambda_2\Lambda_3\Lambda_4}(z_1, z_2) = \frac{1}{\Omega_{\text{BZ}}}\int_{\text{BZ}} d^3k\,\tau_{\Lambda_1\Lambda_2}(\mathbf{k}, z_1)\tau_{\Lambda_3\Lambda_4}(\mathbf{k}, z_2)$$

$$- \tau_{\text{CPA},\Lambda_1\Lambda_2}^{00}(z_1)\tau_{\text{CPA},\Lambda_3\Lambda_4}^{00}(z_2)\,. \qquad (4.87)$$

With Eq. (4.86) one has a closed set of equations which allows the calculation of the terms $\tilde{\sigma}_{\mu\nu}^{0}$ and $\tilde{\sigma}_{\mu\nu}^{1}$ [89]:

$$\tilde{\sigma}_{\mu\nu} = \tilde{\sigma}_{\mu\nu}^{0} + \tilde{\sigma}_{\mu\nu}^{1} \qquad (4.88)$$

$$= -\frac{\hbar}{\pi\Omega}\sum_{\alpha} x_{\alpha}\,\text{Tr}\,\underline{\tilde{J}}^{\alpha\mu}(z_2, z_1)\underline{\tau}_{\text{CPA}}^{00}(z_1)\underline{j}^{\alpha\nu}(z_1, z_2)\underline{\tau}_{\text{CPA}}^{00}(z_2)$$

$$- \frac{\hbar}{\pi\Omega}\sum_{\alpha,\beta}\sum_{\substack{\Lambda_1,\Lambda_2 \\ \Lambda_3,\Lambda_4}} x_{\alpha}x_{\beta}\,\text{Tr}\,\tilde{J}_{\Lambda_1\Lambda_2}^{\alpha\mu}(z_2, z_1)\left[(1 - \chi\omega)^{-1}\chi\right]_{\substack{\Lambda_1\Lambda_2 \\ \Lambda_3\Lambda_4}}\tilde{j}_{\Lambda_3\Lambda_4}^{\beta\nu}(z_1, z_2)\,,$$

$$(4.89)$$

with

$$\tilde{\underline{j}}^{\alpha\mu}(z_1, z_2) = \tilde{\underline{D}}^{0\alpha}(z_1)\underline{j}^{\alpha\mu}(z_1, z_2)\underline{D}^{0\alpha}(z_2) \tag{4.90}$$

$$\tilde{\underline{J}}^{\alpha\mu}(z_2, z_1) = \tilde{\underline{D}}^{0\alpha}(z_2)\underline{J}^{\alpha\mu}(z_2, z_1)\underline{D}^{0\alpha}(z_1) \tag{4.91}$$

where the term $(1 - \chi\omega)^{-1}$ accounts for the vertex corrections to the conductivity. For calculations without vertex corrections this term is replaced by the unity matrix.

The presented scheme for the calculation of the conductivity is derived within the CPA. As discussed in Sec. 4.5 this implies that certain classes of diagrams are not considered in the derivation. In addition, it is not possible to include ordering effects in the atomic lattice site occupation. This shortcoming can be avoided by an employment of the NLCPA instead of CPA. The corresponding scheme for a derivation within the NLCPA is sketched in the following section.

4.6.2 Symmetric Part of $\sigma_{\mu\nu}$ within KKR-NLCPA

In the previous section a scheme for the calculation of the conductivity of an disordered alloy is presented. In this section a similar scheme derived by Tulip et al. [92] is presented which is based on the NLCPA.

The starting point for the derivation of an expression for the conductivity within the NLCPA is again the Kubo-Greenwood equation which is rewritten as shown in Eq. (4.66). After this the conductivity is decomposed in a similar way as in Eqs. (4.72-4.73). Due to the fact that the NLCPA is a cluster theory the natural decomposition of the conductivity is to separate the conductivity into an intra-cluster conductivity:

$$\tilde{\sigma}_{\mu\nu}^{0,\text{NLCPA}} = -\frac{\hbar}{\pi\Omega} \sum_{N \in C} \text{Tr} \left\langle \underline{J}^{M\mu}(z_2, z_1)\underline{\tau}^{MN}(z_1)\underline{j}^{N\nu}(z_1, z_2)\underline{\tau}^{NM}(z_2) \right\rangle_c,$$

$$\tag{4.92}$$

where M, N are sites in reference cluster C and an inter-cluster conductivity:

$$\tilde{\sigma}_{\mu\nu}^{1,\text{NLCPA}} = -\frac{\hbar}{\pi\Omega}$$

$$\sum_{C' \neq C} \sum_{N \in C'} \text{Tr} \left\langle \underline{J}^{M\mu}(z_2, z_1)\underline{\tau}^{M,C'+N}(z_1)\underline{j}^{N\nu}(z_1, z_2)\underline{\tau}^{C'+N,M}(z_2) \right\rangle_c,$$

$$\tag{4.93}$$

which describes the conductivity between reference cluster C and all surrounding clusters C'.

The strategy for the further procedure is very similar to the previous section: defining response functions like K and L and searching for a closed set of equations which allow within the NLCPA the calculation of the conductivity including vertex corrections. Finally, one arrives at the following equations [92]:

$$\tilde{\sigma}_{\mu\nu}^{\text{NLCPA}} = \tilde{\sigma}_{\mu\nu}^{0,\text{NLCPA}} + \tilde{\sigma}_{\mu\nu}^{1,\text{NLCPA}} \tag{4.94}$$

$$= -\frac{\hbar}{\pi\Omega} \sum_{\gamma_C} \sum_{N,K,L} P_{\gamma_C} \operatorname{Tr}$$

$$\underline{\tilde{J}}_{\gamma_C}^{KL\mu}(z_2, z_1) \underline{\tau}_{\text{NLCPA}}^{LN}(z_1) \underline{\tilde{j}}_{\gamma_C}^{N\nu}(z_1, z_2) \underline{\tau}_{\text{NLCPA}}^{NK}(z_2)$$

$$-\frac{\hbar}{\pi\Omega} \sum_{\gamma_C,\gamma_{C'}} \sum_{K,L,M,N} \sum_{\substack{\Lambda_1,\Lambda_2 \\ \Lambda_3,\Lambda_4}} P_{\gamma_C} P_{\gamma_{C'}} \operatorname{Tr}$$

$$\tilde{J}_{\Lambda_1\Lambda_2,\gamma_C}^{KL\mu}(z_2, z_1) \left[(1 - \chi\,\omega)^{-1}\chi \right]_{\Lambda_1\Lambda_2\Lambda_3\Lambda_4}^{LMNK} \tilde{j}_{\Lambda_3\Lambda_4,\gamma_{C'}}^{MN\mu}(z_1, z_2) , \tag{4.95}$$

with

$$\underline{\tilde{j}}_{\gamma_C}^{KL\mu}(z_1, z_2) = \sum_N \underline{D}_{KN}^{\dagger,\gamma_C}(z_1) \underline{j}_{\gamma_C}^{N\nu}(z_1, z_2) \underline{D}_{NL}^{\gamma_C}(z_1) \tag{4.96}$$

$$\underline{\tilde{J}}_{\gamma_C}^{KL\mu}(z_2, z_1) = \sum_N \underline{D}_{KN}^{\dagger,\gamma_C}(z_2) \underline{J}_{\gamma_C}^{N\nu}(z_2, z_1) \underline{D}_{NL}^{\gamma_C}(z_1) \tag{4.97}$$

and $\underline{j}_{\gamma_C}^{N\nu}$ ($\underline{j}_{\gamma_C}^{N\nu}$) is a matrix element as given by Eq. (4.69) (Eq. (4.70)) for site N within a cluster with configuration γ_C (P_{γ_C} is the probability for occurrence of a cluster with configuration γ_C with $\sum_{\gamma_C} P_{\gamma_C} = 1$). The matrices $\underline{D}_{NL}^{\gamma_C}$ and $\underline{D}_{NL}^{\dagger,\gamma_C}$ correspond to their CPA-related counterparts given in Eqs. (4.82) and (4.83), respectively. Similar to the CPA case the term $(1-\chi\,\omega)^{-1}$ in Eq. (4.95) accounts for the vertex corrections to the conductivity. The matrices ω and χ are cluster generalizations of the definitions given in the previous section:

$$\omega_{\Lambda_1,\Lambda_2,\Lambda_3,\Lambda_4}^{LMNK} = \sum_{\gamma_C} P_{\gamma_C} \, x_{\Lambda_4\Lambda_1}^{KL} x_{\Lambda_2\Lambda_3}^{MN} \tag{4.98}$$

$$\chi_{\Lambda_1,\Lambda_2,\Lambda_3,\Lambda_4}^{LMNK} = N_c \sum_{C'\neq C} \tau_{\text{NLCPA},\Lambda_1\Lambda_2}^{L,C'+M} \tau_{\text{NLCPA},\Lambda_3\Lambda_4}^{C'+N,K} \tag{4.99}$$

$$= N_c \sum_{C'} \tau_{\text{NLCPA},\Lambda_1\Lambda_2}^{L,C'+M} \tau_{\text{NLCPA},\Lambda_3\Lambda_4}^{C'+N,K} - N_c \tau_{\text{NLCPA},\Lambda_1\Lambda_2}^{LM} \tau_{\text{NLCPA},\Lambda_3\Lambda_4}^{NK} \tag{4.100}$$

$$= \frac{1}{\Omega_{\text{BZ}}} \int_{\Omega_{\text{BZ}}} d^3k \, \tau_{\Lambda_1\Lambda_2}(\mathbf{k}) \tau_{\Lambda_3\Lambda_4}(\mathbf{k}) e^{i\mathbf{k}(\mathbf{R}_L - \mathbf{R}_M + \mathbf{R}_N - \mathbf{R}_K)}$$
$$- N_c \tau_{\text{NLCPA},\Lambda_1\Lambda_2}^{LM} \tau_{\text{NLCPA},\Lambda_3\Lambda_4}^{NK} , \tag{4.101}$$

where N_c denotes the number of sites within a cluster (the matrix $\underline{x}^{KL}$ is also a cluster generalization of the corresponding CPA matrix shown in Eq. (4.84)).

Applying the coarse graining procedure described in Sec. 3.2.3 the expression for χ can be rewritten in the following form:

$$\chi^{LMNK}_{\Lambda_1,\Lambda_2,\Lambda_3,\Lambda_4} = \frac{N_c}{\Omega_{BZ}} \sum_{\mathbf{K}_n} \int_{\Omega_{\mathbf{K}_n}} d^3k \, \tau_{\Lambda_1\Lambda_2}(\mathbf{K}_n,\mathbf{k})\tau_{\Lambda_3\Lambda_4}(\mathbf{K}_n,\mathbf{k})e^{i\mathbf{K}_n(\mathbf{R}_L-\mathbf{R}_M+\mathbf{R}_N-\mathbf{R}_K)}$$
$$- N_c\tau^{LM}_{\text{NLCPA},\Lambda_1\Lambda_2}\tau^{NK}_{\text{NLCPA},\Lambda_3\Lambda_4} . \tag{4.102}$$

The most time consuming step is the calculation of the vertex corrections $(1-\chi\omega)^{-1}$ due to the inversion of the super matrix $(1-\chi\omega)$. This super matrix has a dimension $N_c \times N_c \times N_\Lambda \times N_\Lambda$ (N_Λ denotes the number of angular momentum quantum numbers) which gives in the case of the smallest possible fcc cluster ($N_c = 4$) with $L_{\max} = 3$ ($N_\Lambda = 32$) a 16384×16384 matrix.

4.6.3 Anti-Symmetric Part of $\sigma_{\mu\nu}$ within KKR-CPA

In order to calculate the anti-symmetric part of the conductivity tensor via the Kubo-Středa equation one has to evaluate the first and third term given in Eq. (4.61):

$$\sigma^{\text{anti}}_{\mu\nu} = \underbrace{\frac{i\hbar}{2\pi V}\text{Tr}\left\langle \left[\hat{J}_\mu\Im G^+\hat{j}_\nu - \hat{j}_\nu\Im G^+\hat{J}_\mu\right]\Re G^+\right\rangle_c}_{\sigma^{\text{anti},A}_{\mu\nu}=\sigma^{\text{anti},A_{\mu\nu}}_{\mu\nu}-\sigma^{\text{anti},A_{\nu\mu}}_{\mu\nu}}$$
$$+ \underbrace{\frac{e}{2\pi V}\text{Tr}\left\langle \Im G^+(\hat{r}_\mu\hat{J}_\nu - \hat{r}_\nu\hat{J}_\mu)\right\rangle_c}_{\sigma^{\text{anti},B}_{\mu\nu}} , \tag{4.103}$$

with the first term giving the dominant contribution (this is demonstrated in Sec. 7.2 for the anomalous Hall effect and in Sec. 7.1 for the spin Hall effect). If one compares the Kubo-Greenwood equation with these two terms it turns out that again an averaging procedure over two or one Green's function has to be done, respectively. The averaging over a single Green's function can be done with the standard CPA algorithm (therefore $\sigma^{\text{anti},B}_{\mu\nu}$ is not further discussed) whereas the averaging over two Green's functions takes into account the vertex corrections to the conductivity and need a similar treatment as for the symmetric part of $\sigma_{\mu\nu}$.

The conductivity term $\sigma^{\text{anti},A}_{\mu\nu}$

Concerning the first term in Eq. (4.103) the only important difference to the treatment of the Kubo-Greenwood equation is that the real part of the

Green's function is needed in addition to the imaginary part:

$$\Im G^+(\mathbf{r}, \mathbf{r}', E_F) = \Im \sum_{\Lambda \Lambda'} Z_\Lambda^i(\mathbf{r}, E_F^+)\tau_{\Lambda \Lambda'}^{ij}(E_F^+)Z_{\Lambda'}^{j \times}(\mathbf{r}', E_F^+) \tag{4.104}$$

$$\Re G^+(\mathbf{r}, \mathbf{r}', E_F) = \Re \sum_{\Lambda \Lambda'} Z_\Lambda^i(\mathbf{r}, E_F^+)\tau_{\Lambda \Lambda'}^{ij}(E_F^+)Z_{\Lambda'}^{j \times}(\mathbf{r}', E_F^+)$$

$$- \sum_\Lambda \left[Z_\Lambda^i(\mathbf{r}, E_F^+)J_\Lambda^{i \times}(\mathbf{r}', E_F^+)\Theta(r' - r) \right.$$

$$\left. + J_\Lambda^i(\mathbf{r}, E_F^+)Z_\Lambda^{i \times}(\mathbf{r}', E_F^+)\Theta(r - r') \right] \delta_{ij} . \tag{4.105}$$

Using the identities $\Im G^+ = \frac{1}{2i}(G^+ - G^-)$ and $\Re G^+ = \frac{1}{2}(G^+ + G^-)$ one can rewrite $\sigma_{\mu\nu}^{\text{anti},A_{\mu\nu}}$ in the following way:

$$\sigma_{\mu\nu}^{\text{anti},A_{\mu\nu}} = \frac{i\hbar}{2\pi V}\text{Tr}\left\langle \hat{J}_\mu \Im G^+ \hat{j}_\nu \Re G^+ \right\rangle_c \tag{4.106}$$

$$= \frac{1}{4i}\left[\tilde{\sigma}_{\mu\nu}^{\text{anti}}(G^+, G^+) - \tilde{\sigma}_{\mu\nu}^{\text{anti}}(G^-, G^-) \right.$$

$$\left. + \tilde{\sigma}_{\mu\nu}^{\text{anti}}(G^+, G^-) - \tilde{\sigma}_{\mu\nu}^{\text{anti}}(G^-, G^+) \right] , \tag{4.107}$$

with

$$\tilde{\sigma}_{\mu\nu}^{\text{anti}}(G^\pm, G^\pm) = \frac{i\hbar}{2\pi V}\text{Tr}\left\langle \hat{J}_\mu G^\pm(E_F)\hat{j}_\mu G^\pm(E_F) \right\rangle_c \tag{4.108}$$

(the procedure for the second term $\sigma_{\mu\nu}^{\text{anti},A_{\nu\mu}}$ is identical).

Eq. (4.108) is similar to Eq. (4.67) which appears in the calculation of the symmetric part of the conductivity tensor. However, the important difference compared to the calculation of the symmetric part is that the full Green's function is needed. For the calculation of the symmetric part it turns out that all terms which include the second term of Eq. (4.105) (this term is purely real) drop out. Therefore, the second term of Eq. (4.105) can be neglected for the calculation of the symmetric part [89] (this is what one would expect because in the Kubo-Greenwood equation only $\Im G^+$ appears).

Splitting of $\tilde{\sigma}_{\mu\nu}^{\text{anti}}$ into an on-site term $\tilde{\sigma}_{\mu\nu}^{\text{anti},0}$ and an off-site term $\tilde{\sigma}_{\mu\nu}^{\text{anti},1}$ (as shown in Sec. 4.6.1 for $\tilde{\sigma}_{\mu\nu}$) shows that the second term from Eq. (4.105) gives only a contribution to the on-site conductivity. From physical reasons and also from the experience with the symmetric part of the conductivity tensor the on-site contribution is negligible compared to the off-site conductivity. In order to avoid numerical difficulties with the irregular solutions which appear in the second term from Eq. (4.105) the on-site conductivity $\tilde{\sigma}_{\mu\nu}^{\text{anti},0}$ is neglected for the anti-symmetric part of the conductivity tensor.

The calculation of $\tilde{\sigma}_{\mu\nu}^{\text{anti},1}$ can be done with the scheme presented in Sec. 4.6.1. Therefore, the calculation of $\sigma_{\mu\nu}^{\text{anti},A}$ is based essentially on a similar procedure as for the symmetric part which account in a proper way for the vertex corrections.

Chapter 5

Spin Resolved Conductivity

During the last years research activities in spintronics increased very rapidly. The reason for the growing interest in this field is based on the close connection with fundamental scientific questions as well as its impact on technology [5, 8]. Compared to standard electronics where only the charge of the electrons is used, spintronics uses the charge of the electrons in combination with the spin degree of freedom. Fig. 5.1 shows schematically the dependence of the electric current on the spin orientation. The green spin-down electrons which have primarily d character are "stronger" scattered than the the red spin-up electrons. This leads for this schematic example to an electric current dominated by spin-up electrons. The observation that transport properties can be very sensitive to the spin orientation of the electrons is the basis of spintronics.

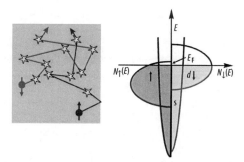

Figure 5.1: Schematic picture of spin dependent electric transport (from Ref. [93]).

One of the most exciting effects within spintronics is the spin Hall effect (SHE) which is discussed in Sec. 7.1. The SHE appears when an electric current flows through a medium with spin-orbit coupling present, leading to a spin current perpendicular to the charge current. For a theoretical investigation of effects like the SHE it is obviously crucial to have a reliable description for the spin-dependent transport that accounts for the impact of spin-orbit coupling in a proper way.

Within non-relativistic quantum mechanics the electronic spin can be described via the well known Pauli matrices σ_i, specifying the non-relativistic spin operator $\mathbf{s} = \frac{\hbar}{2}\boldsymbol{\sigma}$. Due to the fact that the Schrödinger Hamiltonian H_S commutes with $\mathbf{s}$ the projection of the spin e.g. to the z-axis is a constant of motion. This is no longer the case within a scheme that accounts for spin-orbit coupling. The most reliable approach in this context makes use of electronic structure calculations on the basis of the Dirac equation. It turns out that even in the simplest case of a free electron the Dirac Hamiltonian does not commute with e.g. s_z. However, it is possible to define a generalized spin operator which commutes with the free electron Dirac Hamiltonian and shows all characteristic properties of a spin operator (see Sec. 5.2).

Within the fully relativistic description it is not possible to decompose the conductivity in a strict sense into spin-up and spin-down contributions in a simple way. Therefore, one may use approximative schemes or one can switch to scalar-relativistic calculations [94, 95] to decompose the conductivity into two different spin channels. The disadvantage of these two approaches is that approximative schemes work only under certain circumstances and scalar-relativistic calculations neglect all scattering events that lead to a spin flip due to the fact that such calculations neglect spin-orbit coupling. To avoid such shortcomings a proper relativistic spin projection operator is necessary. Most investigations concerning spin-dependent transport within a medium with spin-orbit coupling present were based on the Pauli equation including spin-orbit coupling explicitly as a relativistic correction term and representing the spin-current density essentially by a combination of the Pauli spin matrix σ_z with the conventional current density operator [96]. Very few investigations have been done so far on the basis of the Dirac equation using an expression for the spin-current density, albeit introduced in an ad-hoc manner [84]. In contrast to these approximate schemes to deal with spin-dependent transport the approach suggested by Vernes et al. [97] supplies a fully relativistic and coherent description of electronic spin-polarization and the associated spin-current density. This approach based on the four-component polarization operator $\mathcal{T}$ introduced by Bargmann and Wigner [98] leads, in particular, to a corresponding set of continuity equations (in Sec. 7.1 a spin-current density operator derived from $\mathcal{T}$ will be used in order to calculate the spin Hall effect for non-magnetic alloys). In the following, spin projection operators derived from the polarization operator $\mathcal{T}$ are introduced. This allows a decomposition of the conductivity into contributions from each spin channel within fully relativistic transport calculations. Applications on the spin-dependent transport of various magnetic transition metal alloy systems demonstrate the flexibility and reliability of the new approach.

5.1 An Approximative Spin Decomposition Scheme

In order to decompose the conductivity into two different spin channels one can manipulate certain matrix elements which appear in the Kubo formalism presented in chapter 4 which leads to an approximate spin decomposition scheme. Such a scheme was recently suggested by Popescu et al. [99]. Within the Kubo formalism the various contributions e.g. to σ_{xy} are of the form (see Sec. 4.6):

$$\sigma_{xy} \propto \mathrm{Tr} \left\langle \underline{\underline{j^x}} \, \underline{\underline{\tau}} \, \underline{\underline{j^y}} \, \underline{\underline{\tau}} \right\rangle , \qquad (5.1)$$

where the double underlines indicate a supermatrix with respect to the site and angular momentum indices and $j_{\Lambda\Lambda'}$ is a current density matrix element. This equation can be transformed from the standard relativistic representation (using the quantum numbers $\Lambda = (\kappa, \mu)$ as labels) to a spin-projected one (using the quantum numbers $L = (l, m_l, m_s)$ as labels). Suppressing the spin-flip term of the current density matrix elements $j_{LL'}$ one can easily split the conductivity into spin-up and spin-down contributions and an additional spin-flip contribution σ^{z+-} that is related to the spin-off-diagonal elements of the scattering path operator τ:

$$\sigma_{xy} \propto \mathrm{Tr} \left\langle \begin{pmatrix} \underline{J^x_{++}} & 0 \\ 0 & \underline{J^x_{--}} \end{pmatrix} \begin{pmatrix} \underline{\tau^{nm}_{++}} & \underline{\tau^{nm}_{+-}} \\ \underline{\tau^{nm}_{-+}} & \underline{\tau^{nm}_{--}} \end{pmatrix} \begin{pmatrix} \underline{J^y_{++}} & 0 \\ 0 & \underline{J^y_{--}} \end{pmatrix} \begin{pmatrix} \underline{\tau^{mn}_{++}} & \underline{\tau^{mn}_{+-}} \\ \underline{\tau^{mn}_{-+}} & \underline{\tau^{mn}_{--}} \end{pmatrix} \right\rangle$$
$$(5.2)$$

with $+$=spin-up and $-$=spin-down. From Eq. (5.2) on can easily calculate the contributions of the different spin channels to the conductivity. E.g. the spin-up contribution to the conductivity is:

$$\sigma^{z+}_{xy} \propto \mathrm{Tr} \left\langle \begin{pmatrix} \underline{J^x_{++}} & 0 \\ 0 & 0 \end{pmatrix} \begin{pmatrix} \underline{\tau^{nm}_{++}} & \underline{\tau^{nm}_{+-}} \\ \underline{\tau^{nm}_{-+}} & \underline{\tau^{nm}_{--}} \end{pmatrix} \begin{pmatrix} \underline{J^y_{++}} & 0 \\ 0 & 0 \end{pmatrix} \begin{pmatrix} \underline{\tau^{mn}_{++}} & \underline{\tau^{mn}_{+-}} \\ \underline{\tau^{mn}_{-+}} & \underline{\tau^{mn}_{--}} \end{pmatrix} \right\rangle .$$
$$(5.3)$$

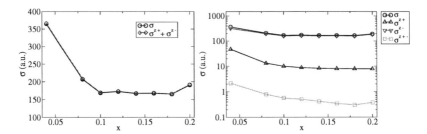

Figure 5.2: Spin decomposition of the $Fe_{1-x}Cr_x$ conductivity. The left figure compares the isotropic conductivity ($\sigma = \frac{2}{3}\sigma_{xx} + \frac{1}{3}\sigma_{zz}$) calculated without spin decomposition scheme with the conductivity calculated via $\sigma^{z+} + \sigma^{z-}$. The right figure shows the spin decomposition of the conductivity.

As an example Fig. 5.2 shows the application of the presented spin decomposition scheme for $Fe_{1-x}Cr_x$. The left panel shows that the deviations between the conductivity calculated via Eq. (5.2) i.e. $\sigma_{xy} = \sigma_{xy}^{z+} + \sigma_{xy}^{z-}$ and the conductivity calculated without spin decomposition are negligible for this system. This demonstrates that the approximate spin decomposition scheme is suitable for $Fe_{1-x}Cr_x$. The right panel of Fig. 5.2 displays the decomposition into spin-up and spin-down contributions. In addition, the small spin-flip conductivity σ^{z+-} is shown. All applications of the scheme made so far imply that it works well for systems containing relatively light elements (i.e. including $3d$ transition metals), but fails if heavier elements are involved (see below).

The spin decomposition scheme shown in this section is based on drastic matrix element manipulations and is therefore only applicable under certain circumstances. If the spin-flip contributions are non-negligible the scheme will fail. Therefore, the need for a more general applicable spin decomposition scheme is obvious. Such a general scheme is presented in the following section.

5.2   Relativistic Spin Projection Operators

The starting point of the derivation of suitable relativistic spin projection operators is based on the four-vector polarization operator $\mathcal{T}$ which was derived by Bargmann and Wigner [98]:

$$\mathbf{T} = \beta\boldsymbol{\Sigma} - \frac{\gamma_5\boldsymbol{\Pi}}{mc} \tag{5.4}$$

$$T_4 = i\frac{\boldsymbol{\Sigma}\cdot\boldsymbol{\Pi}}{mc} , \tag{5.5}$$

with the kinetic momentum $\boldsymbol{\Pi} = (\hat{\mathbf{p}} + \frac{|e|}{c}\mathbf{A})\mathbb{1}_4$ and the canonical momentum $\hat{\mathbf{p}}$. The matrices $\boldsymbol{\Sigma}$ are the relativistic Pauli-matrices, β is one of the standard Dirac matrices and [31]:

$$\gamma_5 = \begin{pmatrix} 0 & -\mathbb{1}_2 \\ -\mathbb{1}_2 & 0 \end{pmatrix} . \tag{5.6}$$

The operator $\mathcal{T}$ can be considered as a generalized spin operator which commutes with the field free Dirac Hamiltonian [31]:

$$H^{\text{free}} = c\boldsymbol{\alpha}\cdot\hat{\mathbf{p}} + \beta mc^2 . \tag{5.7}$$

In addition, the components $\mathcal{T}_\mu$ are the generators of the little group that is a subgroup of the group of Lorentz transformations [100]. In comparison to other suggested forms of polarization operators the operator $\mathcal{T}$ is gauge invariant [31] and therefore the appropriate basis for calculations which include electromagnetic fields.

A widely used relativistic scheme to deal with magnetic solids within spin density functional theory was introduced by MacDonald and Vosko [34]. The corresponding Dirac Hamiltonian (see Eq. (2.17)):

$$H = c\boldsymbol{\alpha}\cdot\hat{\mathbf{p}} + \beta mc^2 + v_{\text{eff}} + \beta\boldsymbol{\Sigma}\cdot\mathbf{B}_{\text{eff}} , \tag{5.8}$$

includes an effective scalar potential v_{eff} and an effective magnetic field $\mathbf{B}_{\text{eff}}$ coupling only to the spin degree of freedom. For the subsequent discussion we choose $\mathbf{B}_{\text{eff}} = B(r)\,\hat{\mathbf{e}}_z$ as frequently done within electronic structure calculations. The commutator of $\mathcal{T}$ and H is non zero which shows that $\mathcal{T}$ is no longer a constant of motion.

From $\mathcal{T}$ corresponding spin projection operators $\mathcal{P}^\pm$ can be derived by demanding:

$$\mathcal{P}^+ + \mathcal{P}^- = 1 \tag{5.9}$$
$$\mathcal{P}^+ - \mathcal{P}^- = \mathcal{T} , \tag{5.10}$$

or equivalently

$$\mathcal{P}^\pm = \frac{1}{2}(1 \pm \mathcal{T}) . \tag{5.11}$$

The projection of $\mathcal{T}$ to a unit vector along the z-axis $\mathbf{n}^T = (0, 0, 1, 0)$ leads to the following expression:

$$\mathcal{T} \cdot \mathbf{n} = \beta \Sigma_z - \frac{\gamma_5 \Pi_z}{mc} \ . \tag{5.12}$$

Making use of the relation $\mathbf{B} = \nabla \times \mathbf{A}$ between the vector potential $\mathbf{A}$ and the magnetic field $\mathbf{B}$, $\mathbf{A}$ has only non-zero components in the xy-plane if $\mathbf{B} \parallel \hat{\mathbf{e}}_z$ (see Eq. (5.8)), i.e. $A_z = 0$. For the spin projection operators this leads to:

$$\mathcal{P}_z^{\pm} = \frac{1}{2} \left[1 \pm \left(\beta \Sigma_z - \frac{\gamma_5 \hat{p}_z}{mc} \right) \right] \ . \tag{5.13}$$

Starting from the polarization operator $\mathcal{T}$ Vernes et al. [97] could demonstrate that a corresponding spin current density operator is given by a combination of $\mathcal{T}$ with the conventional relativistic electron current density operator:

$$\hat{j}_\mu = -ec \, \boldsymbol{\alpha}_\mu \ , \tag{5.14}$$

where $\boldsymbol{\alpha}_\mu$ is one of the standard Dirac matrices [31] ($e = |e|$). Accordingly, we get an operator for the spin-projected current density by combining $\mathcal{P}_z^{\pm}$ and $\hat{j}_\mu$ which leads to $\mathcal{J}_\mu^{z\pm} = \mathcal{P}_z^{\pm} \hat{j}_\mu$.

Using $\mathcal{J}_\mu^{z\pm}$ to represent the observable within Kubo's linear response formalism one can derive expressions for a corresponding spin-projected conductivity tensor (see chapter 4). Restricting to the symmetric part of the tensor one arrives at (see Eq. (4.61)):

$$\sigma_{\mu\nu}^{z\pm} = \frac{\hbar}{\pi N \Omega} \operatorname{Tr} \left\langle \mathcal{J}_\mu^{z\pm} \, \Im G^+(E_F) \, \hat{j}_\nu \, \Im G^+(E_F) \right\rangle \ . \tag{5.15}$$

Here N is the number of atomic sites, Ω the volume per atom, $\hat{j}_\mu$ is the current density operator ($\mu = x, y, z$) and $\Im G^+(E_F)$ is the imaginary part of the retarded one particle Green function at the Fermi energy E_F.

Eq. (5.15) is obviously the counter-part to the conventional Kubo-Greenwood equation [89] for the spin-integrated conductivity that is recovered by replacing $\mathcal{J}_\mu^{z\pm}$ by $\hat{j}_\mu$.

5.3 Results

As a first application of the presented projection scheme the spin resolved conductivity of the alloy system $Fe_{1-x}Cr_x$ has been calculated assuming the magnetization to be aligned along the z-axis. The presence of the spin-orbit coupling gives rise to the anisotropic magnetoresistance (AMR, see Sec. 7.2)

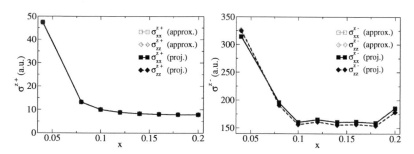

Figure 5.3: Spin resolved conductivity tensor elements $\sigma_{xx}^{z+(-)}$ and $\sigma_{zz}^{z+(-)}$ of $Fe_{1-x}Cr_x$ calculated for the magnetization pointing along the z-axis using the operators $\mathcal{P}_z^{\pm}$ (full symbols). In addition, results are shown that have been obtained using the approximate scheme from Sec. 5.1.

with the conductivity tensor elements $\sigma_{xx} = \sigma_{yy} \neq \sigma_{zz}$ for this situation. The reduced symmetry is also reflected by the spin projected conductivities $\sigma_{xx}^{z+(-)}$ and $\sigma_{zz}^{z+(-)}$, as can be seen in Fig. 5.3. Obviously, the conductivity is quite different for the two spin channels. This behavior can be traced back straight forwardly to the electronic structure of the alloy system around the Fermi energy that can be represented in a most detailed way in terms of the spin-projected Bloch spectral function (BSF) [101]. While for the spin-down subsystem there exists a well-defined Fermi surface with dominant sp-character corresponding to a sharp BSF, the spin-up subsystem is primarily of d-character that is much more influenced by the chemical disorder in the system leading to a BSF with rather washed-out features. As the width of the BSF can be seen as a measure for the inverse of the electronic lifetime the very different width found for the two spin subsystems explain the very different spin-projected conductivities. The influence of an increasing Cr concentration for Fe rich $Fe_{1-x}Cr_x$ alloys on the BSF and the residual resistivity is discussed in detail in Sec. 6.2.2.

Fig. 5.3 shows in addition results that have been obtained on the basis of the approximate spin-projection scheme presented in Sec. 5.1. For $3d$-elements with a relatively low spin-orbit coupling it is found that the neglect of spin-off-diagonal elements of $j_{LL'}$ is well justified and that σ^{z+-} is quite small. In fact the spin-projected conductivities $\sigma_{xx}^{z+(-)}$ and $\sigma_{zz}^{z+(-)}$ obtained by the approximate scheme compare very well with the results using the spin projection operators $\mathcal{P}_z^{\pm}$ (see Fig. 5.3).

In Fig. 5.4 the operators $\mathcal{P}_z^{\pm}$ are applied to the alloy system $Co_{1-x}Pt_x$.

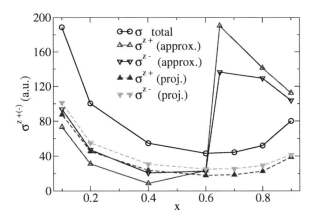

Figure 5.4: Isotropic spin resolved conductivity $\sigma^{z+(-)} = (2\sigma^{z+(-)}_{xx} + \sigma^{z+(-)}_{zz})/3$ of $Co_{1-x}Pt_x$ for the magnetization pointing along the z-axis (full symbols). In addition, results are shown that have been obtained using the approximate scheme from Sec. 5.1.

The results are again compared with calculations using the approximative scheme. This figure clearly shows that the operators $\mathcal{P}^{\pm}_z$ are also suitable for alloys with strong spin-orbit coupling whereas the approximative scheme fails as it leads to unrealistic results. The calculation of the total conductivity for e.g. $Co_{0.2}Pt_{0.8}$ via Eq. (5.2) leads to a conductivity which is approximately six times larger than the conductivity calculated with the Kubo-Greenwood equation without any spin decomposition. This shows that the neglect of spin-off-diagonal elements of $j_{LL'}$ is not justified for this system.

Another important issue that can be seen from Fig. 5.4 is that the difference between the contributions of the two spin channels to the total conductivity is small. This observation is in agreement with the calculations of Ebert et al. [94]. They calculated the ratio between the spin-up and the spin-down resistivity within the two-current model and obtained values between ≈ 1.0-1.8 (present work ≈ 1.0-1.2 using $\mathcal{P}^{\pm}_z$).

As another application of the operators $\mathcal{P}^{\pm}_z$ results for diluted Ni-based alloys with $x_{Ni} = 0.99$ are shown in Fig. 5.5 in terms of the isotropic residual resistivities $\rho^{z+(-)} = ((2\sigma^{z+(-)}_{xx} + \sigma^{z+(-)}_{zz})/3)^{-1}$. As one notes, the resistivity for the two spin channels show a rather different variation with the atomic number of the impurities. This can be traced back again to the spin-projected electronic structure of Ni at the Fermi level and the position of the impurity

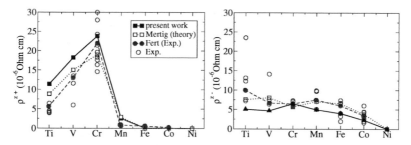

Figure 5.5: Spin resolved resistivity of Ni with $3d$ transition metal impurities (1%) obtained by the present scheme (full squares) compared to theoretical data from Mertig et al. [102] (open squares), experimental data from Fert [103] (full circles) and other experimental data (see Ref. [104], open circles). The left and right panel show the data for spin-up and for spin-down, respectively.

d-states [102]. In Fig. 5.5 the results of calculations by Mertig et al. [102] have been added, that were done in a scalar-relativistic way – i.e. ignoring spin-orbit coupling – on the basis of the Boltzmann-formalism and by making use of the two-current model. In spite of the various differences between this approach and the presented scheme, the resulting spin-projected resistivities agree fairly well. This also holds concerning corresponding experimental data that have been deduced from measurements relying on the two-current model.

In summary, two schemes for a spin projection within transport calculations have been presented. The spin projection operators from Sec. 5.2 have been derived from a relativistic four-vector polarization operator which can be considered as a generalized spin operator which exhibits similar features as the well known non-relativistic spin operator.

The applications presented were restricted to the symmetric part of the corresponding conductivity tensor described by a Kubo-Greenwood-like equation. Results obtained for the disordered alloy systems $Fe_{1-x}Cr_x$, $Co_{1-x}Pt_x$ and diluted Ni-based alloys were compared to results based on an alternative but approximate projection scheme and theoretical as well as experimental data based on the two current model. The good agreement found for the investigated systems ensures the consistency and reliability of the presented spin projection operator scheme. Accordingly, this is expected to hold also when dealing with spin-projected off-diagonal conductivities as e.g. $\sigma_{xy}^{z+(-)}$ on the basis of Kubo-Středa-like equations. This gives access in particular to the spin-projected Hall conductivity in magnetic materials as well as to the spin Hall conductivity in non-magnetic materials which is investigated in Sec. 7.1.

Chapter 6

Residual Resistivity Calculations

6.1 $Ga_{1-x}Mn_xAs$

The system $Ga_{1-x}Mn_xAs$ belongs to the class of so-called diluted magnetic semiconductors (DMS). These systems combine ferromagnetism with semiconducting properties which makes them promising for future spintronics applications [5]. In order to use such materials in technical applications a high Curie temperature T_C (above room temperature) is needed. $Ga_{1-x}Mn_xAs$ belongs to DMS-systems for which Curie temperatures are predicted theoretically above room temperature for Mn concentrations $\geq$ 10% [105]. In spite of these optimistic predictions up to now the highest measured Curie temperatures are $\approx$ 185 K (12.5 % Mn) [106].

The measured Curie temperature is highly affected by the quality of the grown samples [105]. Therefore, the understanding of the influence of the defects appearing during the grow process like e.g. the occupation of interstitial positions by Mn is essential for further applications of these materials. However, such defects play also a crucial role for transport properties like e.g. the residual resistivity of the grown samples. In order to apply $Ga_{1-x}Mn_xAs$ in further spintronics applications a detailed understanding of the transport properties of imperfect samples is required. The present work investigates the influence of several defect types on the residual resistivity of $Ga_{1-x}Mn_xAs$.

The preferred position of Mn in the GaAs host (ZnS structure) is the Ga position ($Mn^{(Ga)}$). In this position Mn acts as an acceptor. In addition, Mn can occupy the less favorable interstitial position ($Mn^{(i)}$) [107] which leads to a double donor behavior for the Mn impurities [105].

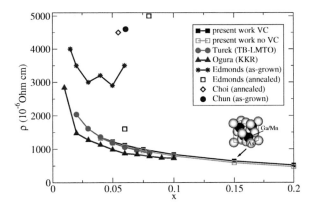

Figure 6.1: Residual resistivity of $Ga_{1-x}Mn_xAs$ as function of the $Mn^{(Ga)}$ concentration. The calculated data of the present work are (VC = vertex corrections) compared with other theoretical works (Turek et al. [108], Ogura unpublished) as well as experimental data (Edmonds et al. [109], Choi et al. [110], Chun et al. [111]).

Fig. 6.1 shows the residual resistivity as calculated in the present work for $Ga_{1-x}Mn_xAs$ as function of the $Mn^{(Ga)}$ concentration. Obviously, the resistivity decreases with increasing $Mn^{(Ga)}$ concentration. This behavior is opposite to that found for metallic alloys which show an increase of resistivity with increasing impurity concentration due to an increase of impurity scattering. Such a metallic like behavior is also present in $Ga_{1-x}Mn_xAs$ but the increased $Mn^{(Ga)}$ concentration leads to an increasing number of carriers which overcompensates the impurity scattering effect.

The agreement of the calculated data with other theoretical works is very good and the observation that vertex corrections only slightly change the resistivity is in line with the calculations by Turek et al. [108] (in the following only results including vertex corrections are shown).

The experimental data shown in Fig. 6.1 show a strong scatter. This observation hinders a direct comparison with calculated resistivities. Nevertheless, all experimental resistivities are above the calculated resistivities. In order to simulate a more realistic sample one has to consider various types of defects. This is illustrated in Fig. 6.2 where the residual resistivity of $Ga_{1-x}Mn_xAs$ samples with additional 1% $Mn^{(i)}$, 1% $As^{(Ga)}$ (As antisites) as well as combinations of 1% $Mn^{(i)}$ with 1% $As^{(Ga)}$ are shown. It turns out that the addition

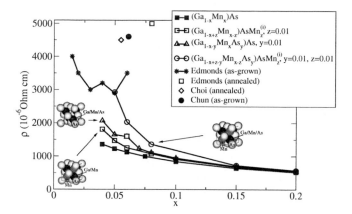

Figure 6.2: Residual resistivity of $Ga_{1-x}Mn_xAs$ with several additional types of defects. The experimental data are similar to Fig. 6.1.

of further types of defects lead to an increase of resistivity. This is based on the fact that $Mn^{(i)}$ as well as $As^{(Ga)}$ act as a double donor and therefore compensate the holes created by two $Mn^{(Ga)}$ [112] leading to a reduction of carriers and therefore to an increased resistivity. The combination of $Mn^{(i)}$ with $As^{(Ga)}$ leads, as expected, to the highest resistivity values and this way makes contact to experimental resistivities. It is noteworthy that the resistivity of $Ga_{0.95}Mn_{0.05}As$ compared to the resistivity of $(Ga_{0.95}Mn_{0.04}As_{0.01})AsMn^{(i)}_{0.01}$ is increased by $\approx 300\%$ which clearly indicates the sensitivity of the system to imperfections. This sensitivity is obviously responsible for the large scatter in the experimental data. Therefore, the comparison with experimental data remains difficult due to uncontrolled impurity creation during the growth process.

6.2 Influence of Short Ranged Ordering

The measurement of the electrical resistivity of a conducting solid is easy to perform and can be used to characterize its microstructure. As many other physical properties of alloys, the residual resistivity is affected by short-range order (SRO) and so its measurement can be used to monitor changes in SRO as they occur in material processing. For technical applications it is important to know what these changes are so that physical properties can be controlled. SRO plays an important role concerning this and resistivity measurements are used to follow its variation [113]. Therefore, the dependence of resistivity on SRO needs to be understood in detail. On intuitive grounds many materials are expected to increase their resistivity with increasing disorder [3] - fluctuations in the occupation of the lattice sites lead to so-called disorder scattering of the current-carrying electrons and thus to an increase in the resistivity. On the same grounds the resistivity is expected to drop if SRO is introduced into the material. Many alloys follow this general trend, e.g. $Cu_{1-x}Zn_x$ [3] (see Sec. 6.2.1). However, there is a significant set of alloys which show completely contrary behavior so that their resistivities actually increase when SRO is increased. These materials belong to the class of so-called K-state alloys [114] which are discussed in detail in Sec. 6.2.3.

In the following three different ordering situations are simulated:

— random disorder (disorder)
    no correlations between occupation of neighboring lattice sites

— SRO
    enhanced probability for unlike atom types sitting next to each other

— clustering
    enhanced probability for like atom types sitting next to each other.

6.2.1 $Cu_{1-x}Zn_x$

In order to demonstrate the applicability of the formalism shown in Sec. 4.6.2 the alloy system $Cu_{1-x}Zn_x$ is a suitable test case. The phase diagram of this alloy system shows a lot of complex phases [115]. However, for approximately equiatomic concentrations the system exhibits above 468 °C a randomly disordered bcc phase which transforms for temperatures below 468 °C to a ordered CsCl structure [116]. Therefore, bcc $Cu_{1-x}Zn_x$ has been used in the following to simulate the impact of ordering effects on the residual resistivity. Randomly disordered bcc $Cu_{1-x}Zn_x$ exhibits an ideal parabolic like $x(1-x)$ behavior of the concentration dependent residual resistivity (Nordheim behavior [117]) as shown by Fig. 6.3. This figure demonstrates in addition the importance of vertex corrections for this system.

As expected in general $Cu_{1-x}Zn_x$ shows a decrease in resistivity upon ordering [3]. For the inclusion of short ranged ordering effects of the lattice site occupation within resistivity calculations the NLCPA formalism introduced in Sec. 4.6.2 has been employed. Short ranged ordering can be defined via

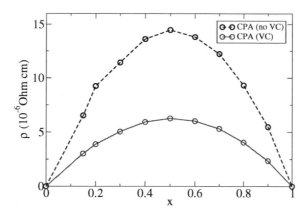

Figure 6.3: Residual resistivity of randomly disordered $Cu_{1-x}Zn_x$. The dashed line indicates the CPA results neglecting vertex corrections (VC) whereas the solid line connects the results including vertex corrections.

the Warren-Cowley SRO parameter α [118, 119]:

$$\alpha_{ij} = 1 - \frac{p_i^A|j^B}{c_A} \qquad (6.1)$$

where c_A is the concentration of atom type A and $p_i^A|j^B$ is the conditional probability that an atom of type A occupies site i when site j is occupied with an atom of type B. If the investigated alloy is randomly disordered $p_i^A|j^B$ is equal c_A which leads to $\alpha_{ij} = 0$ whereas in the case of clustering (phase segregation) $p_i^A|j^B = 0$ and $\alpha_{ij} = 1$.

For the calculations the smallest bcc cluster has been used (2-atomic cluster which leads to four different cluster configurations). The associated cluster probabilities for the different ordering situations are shown in appendix B. The results of these calculations are displayed in Fig. 6.4. As one can see from this figure the agreement of the CPA calculations compared with the NLCPA calculations in the case of randomly disorder is very good which shows that the NLCPA implementation (i.e. algorithm) is reliable.

If one includes clustering effects in the NLCPA calculations the resistivity increases whereas the inclusion of SRO ordering leads to a drastic decrease of resistivity especially in the case of $Cu_{0.5}Zn_{0.5}$. The decrease in resistivity by approaching the concentration $Cu_{0.5}Zn_{0.5}$ is in good agreement with the shown experimental data. For $Cu_{0.5}Zn_{0.5}$ one can define the following cluster

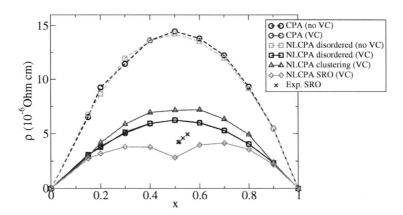

Figure 6.4: Residual resistivity of $Cu_{1-x}Zn_x$ for various ordering situations. In addition, experimental data from Ref. [120] for ordered samples with CsCl structure are displayed.

probabilities:

$$P_{Cu,Cu} = 0.25 + \frac{\alpha}{4}$$
$$P_{Cu,Zn} = 0.25 - \frac{\alpha}{4}$$
$$P_{Zn,Cu} = 0.25 - \frac{\alpha}{4}$$
$$P_{Zn,Zn} = 0.25 + \frac{\alpha}{4} ,$$

where α is the first shell Warren-Cowley SRO parameter which is in the range of 1 (clustering) and -1 (SRO). With these cluster probabilities one can continuously simulate the influence of SRO on the resistivity. Fig. 6.5 compares the resistivity of $Cu_{0.5}Zn_{0.5}$ as function of α with an experimental room temperature measurement. The intersection of the dashed line with the solid line corresponds to $\alpha \approx -0.6$. At room temperature, the atoms form an essentially perfectly ordered CsCl structure [116] what implies $\alpha \approx -1$. The overestimate of α compared to the ordered CsCl structure based presumably on the fact that the calculations where done in the athermal limit (0 K) and therefore all temperature induced scattering effects (e.g. phonons) are neglected which leads to an underestimation of the resistivity compared to experiment. In addition, crystal imperfections lead also to an increase of the

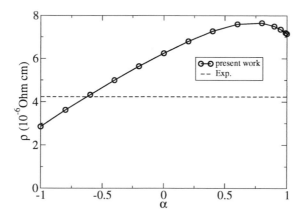

Figure 6.5: Residual resistivity of $Cu_{0.5}Zn_{0.5}$ as function of the first shell Warren-Cowley SRO parameter α. The circles show results including vertex corrections. The dashed line is an experimental value measured at room temperature from Ref. [120].

measured resistivity [3] and therefore complicate the comparison between experimental and theoretical investigations.

This section demonstrates that the inclusion of short ranged ordering effects can drastically influence the residual resistivity. It turns out that relatively small cluster sizes of the NLCPA clusters are already sufficient to simulate the impact of short ranged ordering effects on the residual resistivity.

6.2.2 $Fe_{1-x}Cr_x$ and the Slater-Pauling Curve

The Slater-Pauling plot of average magnetization per atom M versus valence electron number N_v which is shown in Fig. 6.6 plays a pivotal role in the understanding of the properties of ferromagnetic alloys [121]. Its triangular structure of two straight lines with gradients of opposite sign neatly categorizes most alloys into one of two classes where $\frac{dM}{dN_v} = \pm 1$. Long ago Mott [123] pointed out how this behavior can be explained by requiring either the number of majority or minority spin electrons to be fixed. This notion has subsequently been confirmed and given substance by modern spin density functional theory calculations [124–127].

Many DFT calculations for disordered ferromagnetic alloys show that the majority-spin electrons 'see' little disorder and that the majority-spin

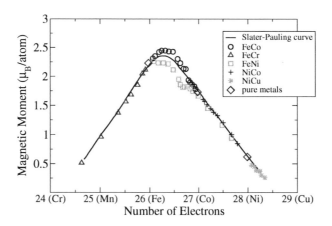

Figure 6.6: The Slater-Pauling curve (from Ref. [122]).

d-states are fully occupied. This leads to $\frac{dM}{dN_v} = -1$ [124]. In contrast the minority-spin electron states are significantly affected by disorder. The overall electronic transport is therefore dominated by the sp-majority-spin electrons. Alloys in this category include fcc-based CoMn, FePt and Ni-rich NiFe alloys.

On the other hand, for some other alloys, typically Fe-rich, bcc-based alloys and many Heusler alloys, the number of minority-spin d-electrons is fixed as the Fermi energy E_F is pinned at a low level in a trough of the d-electron density of states. The property $\frac{dM}{dN_v} = +1$ of the Slater-Pauling curve follows directly from this [124]. It is the ramification of this feature for the electronic transport in such alloys that is investigated here. For the systems considered here disorder is 'seen' strongly by the majority-spin electrons and rather weakly by the minority-spin electrons. Fig. 6.7 provides as an illustration the DOS for bcc $Fe_{0.8}Cr_{0.2}$ disordered alloy. The Fe- and Cr-related minority-spin densities of states curves have similar structure in contrast to those for the majority spin. The Fermi level E_F is positioned in a valley resulting in the average number of minority spin electrons ≈ 3. Moreover from these observations one can expect the resistivity to be dominated by minority spin electrons and to be rather insensitive to overall composition and short-range order. Recently, measurements of the residual electrical resistivity of iron-rich $Fe_{1-x}Cr_x$ alloys have been reported and described as anomalous. The measurements show that the resistivity increases as small amounts of Cr are added to Fe until a plateau is reached ranging from $x = 10\%$ to 20% [128].

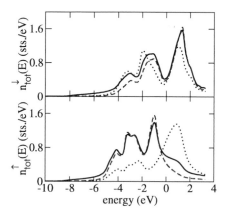

Figure 6.7: Spin projected CPA density of states of disordered bcc $Fe_{0.8}Cr_{0.2}$. The solid line shows the total DOS, the dashed line shows the Fe d-states and the dotted line shows the Cr d-states. The Fermi level is located at the zero of the energy-axis.

This behavior differs markedly from the Nordheim parabolic concentration dependence (see Sec. 6.2.1). An analysis of the data is hindered by a complexity of short-range order in $Fe_{1-x}Cr_x$. Mirebeau et al. [129] reported that for $x < 10\%$ the system develops short-ranged order whereas for larger x short-ranged clustering is found. At higher Cr concentrations still ($> 20\%$), the alloys can undergo on aging a separation into Fe-rich (α) and Cr-rich (α') phases [130, 131] leading either to a miscibility gap or a transformation into a tetragonal σ phase.

The central result of this section is shown in Fig. 6.8. This figure shows the residual resistivity of $Fe_{1-x}Cr_x$ and $Fe_{1-x}V_x$ as a function of the Cr/V concentration x. As mentioned above, the experimental $Fe_{1-x}Cr_x$ data show an anomalous behavior. In the low Cr concentration regime ($\lesssim 10\%$) the residual resistivity increases with increasing Cr concentration. Further increase of the Cr content does not lead to a further increase of the resistivity. The theoretical results show the same variation with the Cr concentration. At 10% Cr the highest value for the resistivity is achieved. If one further increases the Cr concentration, the resistivity stays more or less constant at about 8 $\mu\Omega$cm.

The $Fe_{1-x}V_x$ alloys show a similar behavior for the theoretical residual resistivity. With increasing V concentration the residual resistivity increases up

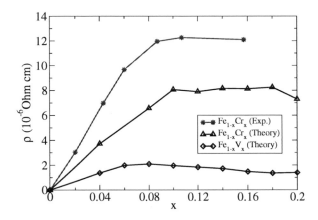

Figure 6.8: The residual resistivity of $Fe_{1-x}Cr_x$ and $Fe_{1-x}V_x$ as a function of the Cr/V concentration x. The asterisks show the experimental data for $Fe_{1-x}Cr_x$ of Ref. [132] at 4.2K. The triangles and diamonds show CPA results for $Fe_{1-x}Cr_x$ and $Fe_{1-x}V_x$, respectively.

to about 2 $\mu\Omega$cm (at 6% V). Further increase of the V concentration leads only to small changes in the residual resistivity.

This behavior can be explained by a different variation of the electronic structure for the majority and minority spin subsystems when the Cr/V concentration changes. Adding Cr to pure Fe in a random way the disorder in the system increases and with this the resistivity increases. This conventional behavior is observed in the regime with a Cr content $\lesssim$ 10% where the system shows a Nordheim like behavior. To identify the contribution of the majority/minority spin subsystems to the conductivity, Bloch spectral functions (BSF, see Sec. 3.2.4) are calculated. Fig. 6.9 shows the total and spin projected BSF for three different Cr concentrations (4%, 12% and 20% Cr). The important observation from the displayed BSF are the different dependencies of the majority and minority spin subsystem on the Cr concentration. At 4% Cr both spin subsystems show sharp peaks for the BSF which indicate that the impact of disorder is relatively small. If one increases the Cr concentration up to 12% a dramatic change occurs. For the BSF of the majority subsystem the prominent lens-shaped band disappears and the remaining rectangular-shaped band become strongly smeared out whereas the minority component is almost unchanged. Further increase of the Cr concentration continues this trend. Due to the fact that an increased broad-

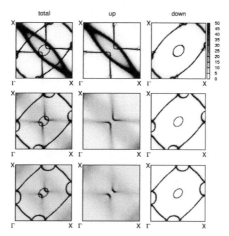

Figure 6.9: Total and spin projected BSF of $Fe_{1-x}Cr_x$ at the Fermi energy in the (001) plane for different Cr concentrations (top: 4% Cr, middle: 12% Cr, bottom: 20% Cr). The black regions correspond to values > 50 a.u.. For a better resolution the cusps of the BSF have been cut.

ening of the BSF leads to a decreased life time of the electronic state, the different behavior of the BSF for the majority spin subsystem compared to the minority spin subsystem indicates that the conductivity is dominated by the minority spin channel.

For a better illustration of the influence of the Cr increase on the minority component Fig. 6.10 shows explicitly the peaks of the BSF at the Fermi energy along the $\Gamma - X$ direction. The BSF shows three main peaks with the last peak (the closest peak to the X point) being split. At 4 % Cr an additional small peak is present to the right of the split peak. This peak is due to a hybridisation with the majority subsystem. This can be inferred from Fig. 6.9 if one compares the minority and majority BSF for 4 % Cr. The reason for this hybridisation is, that in fully relativistic calculations the spin is not a good quantum number because of the presence of spin-orbit coupling [133].

The behavior of the three remaining peaks is quite different. The narrow peak shifts with increasing Cr concentration towards the Γ-point and at 20% Cr this peak overlaps with the peak closest to the Γ-point. The split peak remains nearly fixed at its position but one observes a narrowing with increasing Cr concentration, which corresponds to an increased lifetime of this

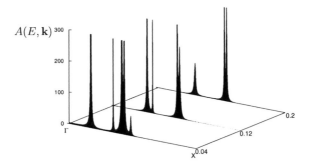

$A(E, \mathbf{k})$

Figure 6.10: BSF along the Γ - X direction for the minority spin subsystem in $Fe_{1-x}Cr_x$ for three different Cr concentrations (4%, 12% and 20% Cr). The cusps of the BSF have been cut at 300 a.u..

state.

For comparison BSF for $Fe_{1-x}V_x$ are also calculated. As expected, the $Fe_{1-x}V_x$ BSF show a similar behavior of the majority/minority spin subsystem as for $Fe_{1-x}Cr_x$. Fig. 6.11 shows the spin projected BSF for $Fe_{0.8}V_{0.2}$. One can see again that the majority component becomes smeared out whereas the minority component displays sharp peaks. To get a more detailed picture of the $Fe_{1-x}V_x$ BSF in Fig. 6.12 a similar picture as shown in Fig. 6.10 for $Fe_{1-x}Cr_x$ is displayed. If one compares Fig. 6.12 with Fig. 6.10 one can see that for $Fe_{1-x}V_x$ the BSF peaks are more sharp than for $Fe_{1-x}Cr_x$. Therefore one can say that the minority spin electrons "see" a smaller difference between Fe and V atoms compared to Fe and Cr atoms. This explains why the residual resistivities are higher in $Fe_{1-x}Cr_x$ compared to $Fe_{1-x}V_x$. Fig. 6.12 shows that the increased disorder due to the increased V concentration does not affect the BSF peaks of the minority spin subsystem. Jen and Chang [134] measured the residual resistivity of polycrystalline $Fe_{1-x}V_x$. They observed a monotonically increase of the residual resistivity in the range of $0.04 < x < 0.2$. They also measured the anisotropic magnetoresistance and obtained a maximum for this quantity at $\approx 7\%$ V. The direct comparison of these measurements with present transport calculations is difficult, however, due to the influence of the polycrystalline nature of the samples on the resistivity that is hard to simulate.

To identify the character of the smeared out states from the majority subsystem, one can project the $Fe_{1-x}Cr_x$ BSF according to its s-, p- and d-contributions. This is shown in Fig. 6.13. The main part of the majority

74

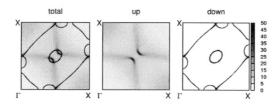

Figure 6.11: Total and spin projected BSF of $Fe_{0.8}V_{0.2}$ at the Fermi energy in the (001) plane. The black regions correspond to values > 50 a.u.. For a better resolution the cusps of the BSF have been cut.

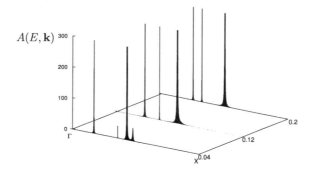

Figure 6.12: BSF along the Γ - X direction for the minority spin subsystem in $Fe_{1-x}V_x$ for three different V concentrations (4%, 12% and 20% V). The cusps of the BSF have been cut at 300 a.u..

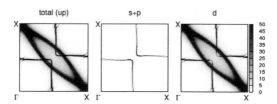

Figure 6.13: Projected majority component of the $Fe_{1-x}Cr_x$ BSF at 4% Cr. The left plot shows the total BSF whereas the middle and the right plot show the $s + p$ and d projected BSF, respectively.

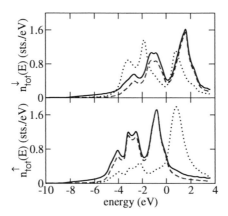

Figure 6.14: Spin projected CPA density of states of $Fe_{0.96}Cr_{0.04}$. The solid line shows the total DOS, the dashed line shows the Fe d-states and the dotted line shows the Cr d-states. The Fermi level is located at the zero of the energy-axis.

states has d-like character. These states obviously strongly broaden with increasing Cr concentration. This behavior is opposite to that of the minority subsystem although this is also dominated by d-like states. The different behavior of the d-states for the two spin subsystems can also be seen in the density of states. In Figs. 6.7 and 6.14 the spin projected DOS of $Fe_{1-x}Cr_x$ for two different Cr concentrations is shown. In addition to the total DOS the d-like part of the DOS is also displayed. The DOS shown in Fig. 6.14 is very close to that of pure Fe. One can see that for the majority component also the antibonding Fe d-states are occupied whereas for the minority component the Fermi level is located in a so-called pseudogap below the antibonding states. Fig. 6.7 clearly shows the relative positions of the Cr and Fe d-states. These states are strongly hybridized for the minority component. The opposite happens in the majority spin channel, where the Fe and the Cr d-states are well separated in energy.

With increasing Cr concentration the antibonding Fe d-peak of the majority component becomes more and more depopulated and new states appear above the Fermi level. Olsson et al. [131] showed that this leads to a completely smeared out band at approximately equiatomic composition. If one compares this with the behavior of the d-states of the minority component, one can see that the increase in Cr concentration has no effect on the total

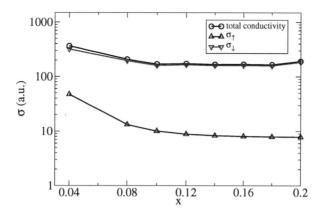

Figure 6.15: Spin resolved conductivities $\sigma_\uparrow$ and $\sigma_\downarrow$ using a logarithmic scale for $Fe_{1-x}Cr_x$ as a function of the concentration x.

DOS of the minority component.

It is well known that the DOS of $Fe_{1-x}V_x$ consists of a minority spin subsystem with the Fermi energy pinned in a pseudogap and a majority spin subsystem which becomes broadened and depopulated with increasing V concentration [125, 127]. This characteristic similar for the $Fe_{1-x}Cr_x$ and $Fe_{1-x}V_x$ DOS is responsible for following the Slater-Pauling curve of the magnetic moment for these alloys. Due to the fact that the number of minority electrons $(N_\downarrow)$ is independent of the Cr/V concentration the magnetization per atom M varies linearly with the Cr/V concentration [125]:

$$M = Z - 2N_{d\downarrow} - 2N_{sp\downarrow}, \qquad (6.2)$$

with Z the number of valence electrons. The number of sp-electrons in the minority spin system $N_{sp\downarrow}$ changes only very little across the $3d$ row [125]). Therefore one can conclude that the Cr/V concentration independent hybridized Fe and Cr/V d-states of the minority spin subsystem are responsible for the appearance of the Slater-Pauling curve and in addition for the apparently anomalous behavior of the residual resistivity of these materials. In order to confirm the observation that the conductivity is dominated by the minority spin channel the spin resolved conductivity is shown in Fig. 6.15. The spin decomposition is achieved via the scheme presented in Sec. 5.1. The results for $\sigma_\uparrow$ and $\sigma_\downarrow$ indeed confirm the picture that evolved from the BSF; i.e. $\sigma_\downarrow$ is about two orders of magnitude larger than $\sigma_\uparrow$ and is nearly concentration independent for $x > 8\%$.

In summary, the resistivity increase from 0-10% Cr (0-6% V) is due to the increased disorder scattering for the majority spin subsystem; roughly speaking, the contribution of the majority subsystem to the conductivity drops down. This drop down can be explained by a smeared out BSF for this component. The increasing Cr/V concentration leads to a broadening of the BSF. This broadening can be related to a decrease of the lifetime of the investigated electronic state which leads to an increasing resistivity. At higher Cr/V concentrations only the minority subsystem contributes to the conductivity. The increase of the Cr/V concentration leads to no broadening of the minority states. Therefore, the contribution of this component to the conductivity in the range of 10-20% Cr (6-20% V) is constant. This leads to a nearly constant resistivity in that concentration regime.

The next step in the analysis of the theoretical results is to investigate the influence of SRO effects on the residual resistivity as this was suggested to be the reason for the anomalous concentration dependence of the residual resistivity. To include SRO effects the NLCPA is employed (see Sec. 3.2.3). Therefore one has to define appropriate cluster configurations and their associated probabilities (P_γ). These probabilities are specified in appendix B. To display the contributions of the different cluster configurations to the density of states of a disordered $Fe_{1-x}Cr_x$ crystal in Fig. 6.16 the cluster resolved density of states for two different Cr concentrations is shown. From this figure one can see that the total DOS agrees very well with the total CPA DOS from Figs. 6.7 and 6.14. The most dominant contribution to the total DOS comes from the FeFe-cluster due to the high Fe concentration. If one compares Fig. 6.16 with Figs. 6.7 and 6.14 one obtains a similar behavior of the minority/majority spin subsystem. The minority part of the DOS shows a hybridization between all cluster configurations whereas the majority part shows a separation in energy between the FeFe/FeCr- and CrCr/CrFe-clusters.

Fig. 6.17 shows the residual resistivity calculated within the NLCPA. In this plot again the three curves from Fig. 6.8 are shown and additionally some curves for different ordering situations. The NLCPA results for disordered $Fe_{1-x}Cr_x$ shows good agreement with the CPA results.

Experimentally it is observed that $Fe_{1-x}Cr_x$ tends for $x < 0.1$ to SRO and for $x > 0.1$ to clustering [129] whereas $Fe_{1-x}V_x$ tends to SRO [135]. Therefore, in Fig. 6.17 NLCPA calculations which simulate SRO ($Fe_{1-x}Cr_x$ and $Fe_{1-x}V_x$) as well as clustering ($Fe_{1-x}Cr_x$) are displayed.

The important observation from the calculations is that the influence of ordering effects (SRO and clustering) on the residual resistivity is very small for these systems. In Sec. 6.2.1 and Sec. 6.2.3 the same formalism (using 2- and

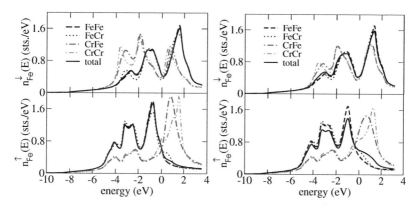

Figure 6.16: Cluster resolved NLCPA density of states of disordered $Fe_{1-x}Cr_x$ (left: 4% Cr, right: 20% Cr). The solid (black) line shows the total DOS, the dashed (red) line shows the contribution of the FeFe-cluster, the (blue) dotted line of the FeCr-cluster, the (green) dashed-dotted line of the CrFe-cluster and the (orange) dashed-dotted-dotted line of the CrCr-cluster.

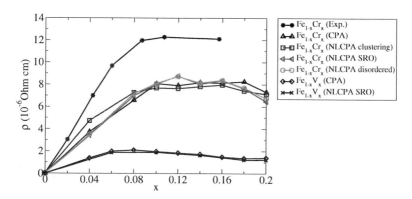

Figure 6.17: The resistivity of $Fe_{1-x}Cr_x$ and $Fe_{1-x}V_x$ for several ordering situations.

4-atomic clusters) is applied to bcc $Cu_{1-x}Zn_x$ and fcc $Ag_{1-x}Pd_x$, respectively and gives a strong variation of the resistivity as function of the ordering state. The small influence of ordering effects on the resistivity for $Fe_{1-x}Cr_x$ was also found experimentally by Mirebeau et al. [129]. This shows that the NLCPA formalism for the calculation of the residual resistivity can handle various different ordering dependences of the residual resistivity. Nevertheless, one has to mention that only small cluster sizes with two atoms are used in the calculations. Therefore, it is not possible to preclude that bigger cluster sizes would have a bigger impact on the transport properties.

In summary, the present work deals on an *ab-inito* level with the anomalous concentration dependence of the residual resistivity in $Fe_{1-x}Cr_x$ and $Fe_{1-x}V_x$ alloys ($x \leq 0.2$). It turns out, that the concentration independence of the residual resistivity can be related to the specific behavior of the electronic structure with increasing alloying which is common for alloys of this branch of the Slater-Pauling plot. Therefore, the behavior of the residual resistivity is by no means anomalous for these materials and one can predict a similar behavior for other alloy systems from the same branch of the Slater-Pauling plot. In addition, the influence of short-ranged correlations in the lattice site occupation by employing the NLCPA formulation of the Kubo-Greenwood equation is investigated. The inclusion of such short-range effects has only little influence on the results.

The comparison of the $Fe_{1-x}Cr_x$ calculations with experiment show satisfying agreement. The difference in the height of the plateau, as compared to the experiment, could be attributed to impurities, lattice defects, grain boundaries, etc. which are always present in samples and therefore influence the experimental data. Such imperfections, which have been neglected in the present calculations, lead in general to an increase of the measured resistivity [3]. The calculations show the initial linear increase albeit with a lower slope than seen in experiment. In Ref. [128] it is argued that this is a consequence of the limitations of the CPA for alloys with a dominant concentration of one constituent. However, it should be pointed out that the NLCPA results confirm the single-site CPA data. The calculations reveal a plateau of the residual resistivity starting at the same Cr concentration as seen in experiment (at 10% Cr). This is in variance to an earlier theoretical study [128] which finds the starting point of the plateau only at $\approx$ 20% Cr.

6.2.3 The K-effect

There is a significant set of alloys which show a completely opposite behavior compared to the observations from Sec. 6.2.1. These alloys show a decreasing resistivity with decreasing ordering. This phenomenon is often discussed in terms of a "komplex" state or K-state [114]. Such alloys are typically rich in late transition metals such as Ni, Pd or Pt and alloyed with a mid-row element such as Cr, Mo or W. This behavior was first observed in $Ni_{1-x}Cr_x$ [114] which was cold-worked what means that the metal is plastically deformed at a temperature low relative to its melting point [136, 137] and later in different materials [138–142] such as $Ni_{1-x}Mo_x$ and $Pd_{1-x}W_x$ alloys. When a metallic alloy is cold-worked the number of dislocations and other defects is increased and atomic SRO is reduced [136]. In an cold-worked alloy solute atoms can congregate in the regions around defects [137].

The K-state group contains some important industrial materials as for example $Ni_{1-x}Cr_x$ "Nichrome" alloys that are well suited for the production of high-quality resistors [143], high temperature devices (e.g. turbine blades in steam engines [144]) and corrosion protected devices [145].

In order to simulate the K-effect the NLCPA medium has been constructed via the smallest possible fcc cluster allowing for non-local correlations over nearest atomic neighbor length scales. For a fcc lattice this implies a cluster of $N_c = 4$ atoms being considered [66, 146]. In a disordered binary fcc alloy, $A_{1-x}B_x$, there are 16 (2^{N_c}) possible arrangements, of the 2 types of atom distributed over the sites of such a cluster. By weighting these configurations appropriately one can incorporate short-ranged ordering or clustering effects. For example, just 2 configurations, 4 A and 4 B atoms, weighted $1-x$ and x, respectively, describe clustering over nearest neighbor scales whilst 4 equally weighted configurations for the different ways 3 A and 1 B atoms can be arranged simulates short-ranged order in a $A_{0.75}B_{0.25}$ alloy. In a $A_{0.80}B_{0.20}$ alloy the configuration weighting to describe a completely random alloy causes a A(B) atom to have on average 9.6 (2.4) like nearest neighbor atoms whereas for our SRO system a A(B) atom has an average of 9.45 (1.8) like nearest neighbors and its counterpart in a short-ranged clustered system has 10.2 (4.8) like nearest neighbors. The explicit cluster probabilities associated with different ordering situations are given in appendix B.

In this section the influence of short-range ordering and clustering on the residual resistivity of $Ni_{0.8}Mo_{0.2}$, $Ni_{0.8}Cr_{0.2}$ and $Pd_{0.8}W_{0.2}$ is presented and analyzed in detail for $Ni_{0.8}Mo_{0.2}$. Furthermore, it will be shown that the same mechanism is valid for cold worked Pd-rich $Pd_{1-x}Ag_x$ alloys.

In Fig. 6.18 the calculated residual resistivities for the alloys $Ni_{0.8}Cr_{0.2}$, $Ni_{0.8}Mo_{0.2}$ and $Pd_{0.8}W_{0.2}$ are shown for three cases: for (i) random disorder, (ii) short-ranged order and (iii) short-ranged clustering using this approach. Comparison of the resistivities calculated within the CPA with the resistivities obtained on the basis of the NLCPA which describe random disorder show good agreement. Nicholson and Brown [142] also investigated $Ni_{0.8}Mo_{0.2}$ and obtained for the disordered case a residual resistivity of $129 \times 10^{-6}\ \Omega cm$

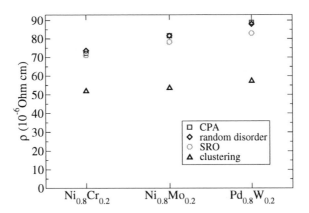

Figure 6.18: The residual resistivities for three investigated alloys $Ni_{0.8}Cr_{0.2}$, $Ni_{0.8}Mo_{0.2}$ and $Pd_{0.8}W_{0.2}$ for random disorder (squares (CPA) and diamonds (NLCPA)), short-ranged order (circles) and short-ranged clustering (triangles).

that is in variance to the present work. The difference is caused presumably by an insufficient angular momentum expansion used in Ref. [142]. For $l_{max} = 2$ the residual resistivity as calculated in this work is 131×10^{-6} Ωcm in perfect agreement with Ref. [142]. An increase of the angular momentum expansion to $l_{max} = 3$ leads to a decrease of the resistivity down to 82×10^{-6} Ωcm ($l_{max} = 4$ gives 81×10^{-6} Ωcm).

It is noteworthy that the resistivities for the clustered configuration are much smaller than those of the other two configurations whilst the resistivities for the disordered and short-ranged ordered configurations are very similar. This shows emphatically how the conductivity increases as the number of like late TM nearest neighbors increases. This effect is consistent with the experimental observation that quenched samples of these materials (in these materials SRO is present according to x-ray experiments [147–149]) show a higher resistivity than the cold-worked samples. The cold-working leads to a decrease of the SRO in the samples and therefore increases the probability of having like atoms sitting next to each other.

A more detailed analysis of the calculated resistivities for $Ni_{0.8}Mo_{0.2}$ is shown in Fig. 6.19. Similar features are found for the other alloys. This figure shows the conductivity decomposed into contributions from different atom pairs. One can clearly see that the conductivity between Ni-Ni pairs

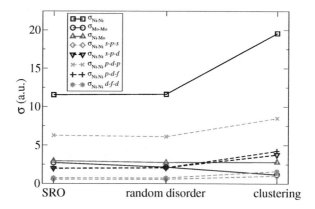

Figure 6.19: Atom type and angular momentum resolved conductivity of $Ni_{0.8}Mo_{0.2}$ for random disorder (center), short-ranged order (left) and short-ranged clustering (right). The squares (black line) show the contribution of Ni-Ni pairs, the circles (blue line) show that of the Mo-Mo pairs and the up-triangles (green line) show the contribution of Ni-Mo pairs to the conductivity. The dashed lines show the decomposition of the Ni-Ni pair conductivity into different scattering channels.

is the dominant contribution which increases significantly when short-ranged clustering is included. This observation is opposite to one of the few theoretical investigations of the K-effect which was directed at $Ni_{0.8}Mo_{0.2}$ alloy [142]. There it was claimed that the smaller resistivity when SRO is reduced arises from an enhanced current flowing between the minority Mo-Mo pairs. Fig. 6.19 also shows the contributions of different scattering channels to the Ni-Ni conductivity accounting for the selection rules [89, 92]. One can see that the dominating scattering channel for this system is the p-d-p-channel. This is unlike typical examples from electronic transport theory where the conductivity is taken to be dominated by the mobile s electrons [3].

One can obtain further insight from the calculations by examining the electronic structure of the alloy and the changes induced by short-ranged ordering or clustering. As can be seen in Fig. 6.20 the DOS for the randomly disordered and the short-range ordered cases are very similar (especially at E_F), whereas for the clustered configuration significant differences appear. If one compares the Ni DOS of short-ranged ordered $Ni_{0.8}Mo_{0.2}$ with the corresponding clustered one it turns out that the increase of the probability

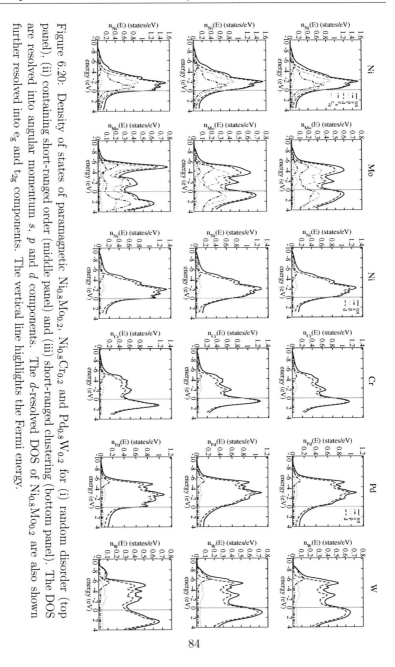

Figure 6.20: Density of states of paramagnetic $Ni_{0.8}Mo_{0.2}$, $Ni_{0.8}Cr_{0.2}$ and $Pd_{0.8}W_{0.2}$ for (i) random disorder (top panel), (ii) containing short-ranged order (middle panel) and (iii) short-ranged clustering (bottom panel). The DOS are resolved into angular momentum s, p and d components. The d-resolved DOS of $Ni_{0.8}Mo_{0.2}$ are also shown further resolved into e_g and t_{2g} components. The vertical line highlights the Fermi energy.

for Ni-Ni neighbors leads to a splitting of the d-state peak into three peaks. There is a weakening of the t_{2g} Ni-Mo bonding states. In particular the new peak close to E_F is caused by an increase of the non-bonding t_{2g} DOS at this energy. The increase of the t_{2g} states leads to an $\approx 40\%$ increase of the d-state density at E_F for the clustered configuration which leads directly to an enhanced conductivity for this configuration. The NLCPA electronic structure calculations directly produce measures of the Bloch spectral function coarse grained over the Brillouin zone [76]. The difference between Bloch spectral function at E_F averaged over a cube centered on the Γ point occupying a quarter of the volume of the Brillouin zone and that of a same sized cube centered on the X-point, $\Delta \bar{A}_B = \bar{A}_B(X) - \bar{A}_B(\Gamma)$, gives a rough average of the Fermi velocity. This quantity varies consistently with the observation that the clustered configuration exhibits the smallest resistivity. For example for $Ni_{0.8}Mo_{0.2}$ $\Delta \bar{A}_B$ becomes -0.66, -0.62 and 8.08 states/atom/$(a.u.)^3$ for the randomly disordered, short-range ordered and short-range clustered cases respectively.

Concluding one can propose the following explanation for the K-effect. In K-state alloys the density of states (DOS) around the Fermi energy E_F is predominantly associated with the d-electrons from the late-row transition metal (TM) element and is large as E_F lies near the top of these d-bands. In a random disordered alloy, or one with SRO, there is however only a modest contribution to current carried by this d-electron channel owing to the low Fermi velocity and short lifetime. For late TMs like Ni, Pd or Pt, the Fermi energy E_F lies in the bands set up by d-orbitals that point between next neighboring atoms. The strong electrostatic repulsion effects associated with these orbitals as described by ligand field theory make these orbitals the last to be occupied. For example in the face centered cubic systems studied in this paper these are the t_{2g} bands - crystal field effects having broken the 5-fold d-degeneracy into e_g and t_{2g} components and the t_{2g} bands lie at higher energy than the e_g ones. When an alloy rich in a late TM is 'worked', the number of unlike nearest neighbor atoms is reduced and the probability for electron hopping between neighboring "like" late TM atoms is enhanced for energies near E_F. It follows that there is an increased bandwidth, larger Fermi velocity and therefore a bigger contribution to the conductivity for this situation.

The studies on $Ni_{1-x}Mo_x$, $Ni_{1-x}Cr_x$ and $Pt_{1-x}W_x$ alloys show that changes of the t_{2g} amplitude and band width for electron hopping between majority TM sites at E_F is responsible for their K-state behavior of the resistivity. It follows that other late TM-rich alloys should behave similarly. In order to demonstrate this the alloy system $Pd_{1-x}Ag_x$ is investigated. This alloy system is well-known for deviation from the Nordheim behavior [3] and a

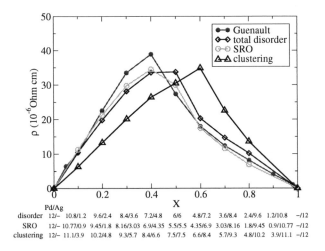

disorder	12/–	10.8/1.2	9.6/2.4	8.4/3.6	7.2/4.8	6/6	4.8/7.2	3.6/8.4	2.4/9.6	1.2/10.8	–/12
SRO	12/–	10.77/0.9	9.45/1.8	8.16/3.03	6.9/4.35	5.5/5.5	4.35/6.9	3.03/8.16	1.8/9.45	0.9/10.77	–/12
clustering	12/–	11.1/3.9	10.2/4.8	9.3/5.7	8.4/6.6	7.5/7.5	6.6/8.4	5.7/9.3	4.8/10.2	3.9/11.1	–/12

Figure 6.21: The resistivity of the $Pd_{1-x}Ag_x$ alloy system for several ordering situations. The experimental results of Guénault [150] are also shown. The panel below shows the number of like nearest neighbors for both a Pd and a Ag atom in each of three configurations.

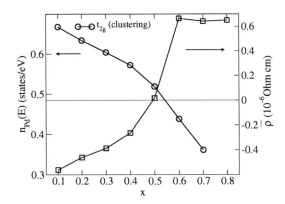

Figure 6.22: The normalized resistivity difference $\bar{\rho}$ of $Pd_{1-x}Ag_x$ compared with the number of Pd t_{2g} states at E_F (for clustering) as function of x. The horizontal line indicates $\bar{\rho} = 0$.

continuous decrease of the d-state density at E_F with increasing Ag concentration. Fig. 6.21 shows the residual resistivity of $Pd_{1-x}Ag_x$ for the three investigated ordering situations and gives the averaged number of like nearest neighbors for each atom type depending on the ordering situation. From this figure one can clearly see the deviation from the Nordheim behavior. Once again one finds that when short-range clustering is simulated for these alloys the calculated resistivity is decreased for the Pd-rich alloys where E_F lies at the top of the t_{2g} states. In order to demonstrate this behavior Fig 6.22 shows the normalized resistivity difference:

$$\bar{\rho} = \frac{\rho_{\text{clus}} - \rho_{\text{ord}}}{\frac{1}{2}(\rho_{\text{clus}} + \rho_{\text{ord}})} \, , \tag{6.3}$$

for $Pd_{1-x}Ag_x$ where ρ_{clus} denotes the resistivity in the case of clustering and ρ_{ord} corresponds to the resistivity when SRO is present. In addition, Fig. 6.22 shows the number of Pd t_{2g} states at E_F as function of Ag concentration x. One can see that with decreasing number of Pd t_{2g} states $\bar{\rho}$ approaches zero and becomes positive for $x \geq 0.5$.

From the investigations of the present work one can conclude that this effect should be present in any late TM-rich alloy which can develop short- and long-ranged order that can be modified by materials processing.

Chapter 6. Residual Resistivity Calculations

Chapter 7

Hall effect

In 1879 Edwin H. Hall observed that a current-carrying non-magnetic conductor exposed to an external magnetic field B shows a transverse voltage [151]. This voltage is based on the Lorentz force which shifts the electrons to one side of the conductor. The corresponding effect is the so-called ordinary Hall effect (OHE). In 1881 Edwin H. Hall reported that he observed in ferromagnetic metals a similar effect which is approximately ten times larger as in non-magnetic conductors [152]. In ferromagnets the Hall voltage consists of two different contributions namely the OHE and the anomalous Hall effect (AHE). The AHE originates from the spin polarization of the carriers in combination with the relativistic spin-orbit interaction and is proportional to the magnetization M of the ferromagnet. Therefore, the AHE is a purely relativistic effect.

These observations lead to the following equation for the Hall resistivity of a ferromagnetic conductor [153]:

$$\rho_{\mathrm{H}} = R_0 B + R_A M \; , \tag{7.1}$$

where R_0 and R_A are the ordinary and anomalous Hall coefficients, respectively.

The spin Hall effect (SHE) was first described 1971 by Dyakonov and Perel [14, 15] and more recently by Hirsch [16]. The SHE appears even in non-magnetic conductors without any external magnetic field. This effect as well as the AHE are based on the observation that e.g. in the case of spin-dependent scattering by impurities "spin-up" electrons are scattered into opposite direction compared to "spin-down" electrons due to spin-orbit coupling [17]. This leads to a spin-up and spin-down accumulation perpendicular to the electric field, respectively. Due to the fact that non-magnetic conductors exhibit no spin-polarization the spin separation shows no accompanying Hall voltage. In the case of the AHE this spin separation leads due to spin

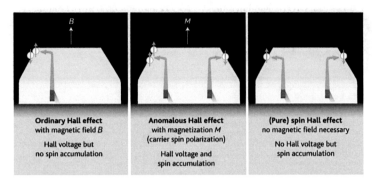

| Ordinary Hall effect with magnetic field *B* | Anomalous Hall effect with magnetization *M* (carrier spin polarization) | (Pure) spin Hall effect no magnetic field necessary |
| Hall voltage but no spin accumulation | Hall voltage and spin accumulation | No Hall voltage but spin accumulation |

Figure 7.1: Schematic illustration of the OHE, AHE and SHE (from Ref. [17]).

imbalance to a charge accumulation which is responsible for the Hall voltage. Despite the difficulties associated with the absence of the Hall voltage in the case of SHE a few groups succeeded in measuring the SHE [13, 154–157]. In order to clarify the differences between the mentioned Hall effects Fig. 7.1 shows in schematic way the OHE, AHE and SHE (in the following only the AHE and especially the SHE are investigated).

The AHE and SHE result from the same microscopical mechanisms. The only difference is that for the AHE spin polarization has to be present in the investigated material. Therefore, the AHE and SHE can be discussed in a very similar way. First of all these effects can be decomposed into two different contributions namely the extrinsic and the intrinsic contribution. The extrinsic contribution to the AHE/SHE is based on spin-dependent scattering e.g. by impurities. Microscopically, two dominant mechanisms are responsible for spin-dependent scattering: skew scattering which results from asymmetric scattering [158, 159] and side-jump scattering [160]. The first mechanism leads to a deflected averaged trajectory of $\approx 0.6°$ wheres the second mechanism corresponds to a lateral displacement ($\approx 10^{-11}$m) of the center of the wave-packet during the scattering process [81]. These two effects are shown schematically in Fig. 7.2.

Crépieux and Bruno [81] identified the corresponding diagrams of the skew and side-jump scattering. They used a diagrammatic representation of the Kubo-Středa equation as shown in Fig. 7.3 (see Sec. 4.5) in combination with a model Hamiltonian calculation. The identified diagrams are shown in Fig. 7.4. Obviously, these diagrams belong to the class of vertex dia-

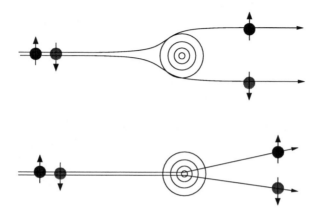

Figure 7.2: Schematic picture of the side-jump (top) and skew scattering (bottom) mechanism which lead to spin-dependent scattering.

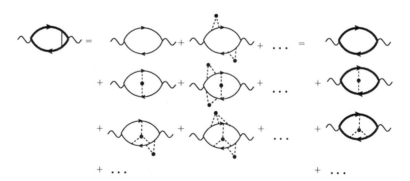

Figure 7.3: Diagrammatic representation of $\sigma_{\mu\nu} = \frac{\hbar V}{2\pi} \mathrm{Tr} \left\langle \hat{j}_\mu \, G^+ \, \hat{j}_\nu \, G^- \right\rangle_c$.

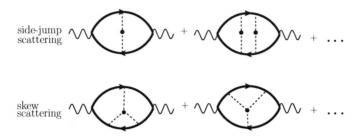

Figure 7.4: Representative diagrams contributing to the side-jump and skew scattering.

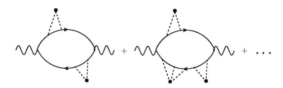

Figure 7.5: Representative non-vertex diagrams contributing to the extrinsic AHE/SHE.

grams. Therefore, the neglect of vertex corrections leads to a neglect of diagrams which are responsible for skew and side-jump scattering. However, non-vertex diagrams as shown in Fig. 7.5 give also a contribution which is impurity based. Nevertheless, in the following only vertex diagrams are discussed as contributions to the extrinsic SHE/AHE whereas all other diagrams are discussed as intrinsic contributions.

The intrinsic AHE/SHE appears in pure systems where no impurity scattering is present. This effect can be related to a geometric phase the so-called Berry phase [83, 153, 161, 162]. In order to calculate the intrinsic AHE/SHE one can use the following equation [82]:

$$\sigma_{xy}^{\text{intr}} = -e^2 \hbar \sum_n \int_{\text{BZ}} \frac{d^3 k}{(2\pi)^3} \, f_n \, \Omega_n(\mathbf{k}) \,, \qquad (7.2)$$

with the Berry curvature:

$$\Omega_n(\mathbf{k}) = -\sum_{n' \neq n} \frac{2\Im \langle \psi_{n\mathbf{k}} | v_x | \psi_{n'\mathbf{k}} \rangle \langle \psi_{n'\mathbf{k}} | v_y | \psi_{n\mathbf{k}} \rangle}{(E_{n'} - E_n)^2} \,, \qquad (7.3)$$

Figure 7.6: Diagrammatic representation of the intrinsic AHE/SHE.

where $|\psi_{n\mathbf{k}}\rangle$ and E_n are the eigenstates and eigenvalues of a Bloch Hamiltonian, respectively, f_n is the equilibrium Fermi-Dirac distribution function and v's are velocity operators. Eq. (7.2) is derived from the Kubo-equation and similar to Eq. (4.30).

Using again the diagrammatic representation of the Kubo-Středa equation one can identify the diagram which corresponds to the intrinsic contribution. This diagram is shown in Fig. 7.6. Due to the fact that the intrinsic contribution is a band structure effect and not related to impurity scattering, the corresponding diagram shows no impurity interactions.

The theoretical investigation of the AHE/SHE is a very challenging research field. In order to calculate the AHE/SHE one needs a transport theory which gives access to the anti-symmetric part of the conductivity tensor in combination with a proper relativistic treatment (inclusion of spin-orbit coupling) of the electronic structure. In addition, one needs a reliable treatment of disorder to determine the extrinsic AHE/SHE contribution and for the calculation of the SHE a spin decomposition scheme has to be applied. Therefore, only few theoretical investigations are available which consider the intrinsic AHE [82, 163, 164] and SHE [84, 165–168] on an *ab initio* level or the intrinsic/extrinsic contribution via model Hamiltonians e.g. [81, 83, 87, 96, 169–171]. The method used in the present work (Kubo-Středa equation in combination with fully relativistic spin polarized KKR and CPA) allows for the first time a simultaneous investigation of the extrinsic as well as intrinsic contribution to the AHE/SHE on an *ab initio* level.

7.1 Spin Hall Effect

The following section investigates the extrinsic and intrinsic spin Hall effect for several non-magnetic $4d$ and $5d$ transition metal alloys. In order to calculate the SHE the Kubo-Středa equation is combined with an appropriate spin-current density operator derived from the Bargmann-Wigner four-vector polarization operator $\mathcal{T}$ (see Sec. 5.2 and appendix E). Vernes et al. [97] derived from this fully relativistic polarization operator a continuity equation and could identify the explicit expression for the spin-current density operator $\hat{J}_i^j = ec\alpha_i T_j$ ($e = |e|$). For a current along the direction of the x-axis and the spin component along the z-axis the spin-current density operator becomes $\hat{J}_x^z = ec\alpha_x T_z$. Taking this operator as representative for the observable one gets the following expression for the spin-current conductivity (see Eq. (4.35)):

$$\sigma_{xy}^z = \frac{\hbar}{2\pi V}\text{Tr}\left\langle \hat{J}_x^z G^+ \hat{j}_y G^- \right\rangle_c + \frac{e}{4\pi iV}\text{Tr}\left\langle (G^+ - G^-)(\hat{r}_x \hat{J}_y^z - \hat{r}_y \hat{J}_x^z) \right\rangle_c, \quad (7.4)$$

where terms containing products of the retarded (advanced) Green's functions only have been dropped as they give only a minor contribution to the spin Hall conductivity (SHC) [162]. It is important to note that the sign of the SHC depends on the definition of the spin-current density operator i.e. using e or $-e$ for the conversion of the spin conductivity into the unit of charge conductivity.

Fig. 7.7 shows the SHC for three different alloy systems: Pt_xIr_{1-x}, Au_xPt_{1-x} and Ag_xAu_{1-x}. In order to avoid numerical instabilities the alloy concentrations range for each alloy system according to $0.01 \leq x \leq 0.99$. In Sec. 4.6.3 it was mentioned that the term $\sigma_{xy}^{\text{anti},Bz}$ from Eq. (4.103) (corresponding to the second term from Eq. (7.4)) has only a negligible influence on the SHC. The contribution of this term to the SHC is displayed in Fig. 7.8. Obviously, the contribution of $\sigma_{xy}^{\text{anti},Bz}$ is indeed negligible compared to the first term from Eq. (7.4).

Fig. 7.7 clearly shows the strong dependence of the role of the vertex corrections (VC) on the alloy system. Vertex corrections to the SHC of Pt_xIr_{1-x} are negligible within the investigated alloy concentration regime whereas Au_xPt_{1-x} and especially Ag_xAu_{1-x} show a dramatic increase of the conductivity when approaching the pure metals if vertex corrections are included. If one subtracts from calculations including vertex corrections the results without vertex corrections one gets the extrinsic SHC contribution. This indicates that the extrinsic SHE is very pronounced in Au with small concentrations of Ag or Pt impurities as well as in Ag with small concentrations of Au impurities. In order to show that the extrinsic contribution can be calculated via the difference between calculations with and without vertex corrections one has to take into account that the conductivity including/neglecting vertex corrections is proportional to:

$$\sigma_{xy}^{\text{VC}} \propto \tilde{J}^x(z_2, z_1)\left[(1 - \chi\omega)^{-1}\chi\right]\tilde{j}^y(z_2, z_1) \quad (7.5)$$

$$\sigma_{xy}^{\text{noVC}} \propto \tilde{J}^x(z_2, z_1)\chi\tilde{j}^y(z_2, z_1) \quad (7.6)$$

(angular momentum indices are neglected, the explicit expression is shown by Eq. (4.89)). The term $(1 - \chi\omega)^{-1}$ accounts for vertex corrections and is

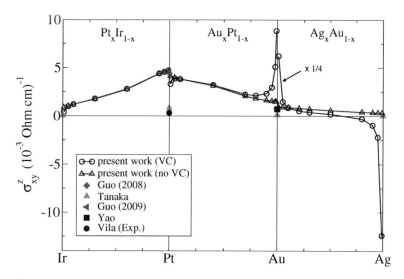

Figure 7.7: The spin Hall conductivity for three different alloy systems: Pt_xIr_{1-x}, Au_xPt_{1-x} and Ag_xAu_{1-x}. The circles (black line) correspond to calculations including vertex corrections (the VC results of Ag_xAu_{1-x} are multiplied by 0.25) and the triangles (blue line) represent calculations without vertex corrections. In addition, data of other theoretical investigations concerning the intrinsic SHE are displayed (Refs. [96, 165, 166, 168]) as well as an experimental data point from Vila et al. [156].

replaced by the unity matrix if vertex corrections are neglected [89]. If one uses:

$$(1 - \chi \omega)^{-1} = 1 + \chi \omega + \chi \omega \chi \omega + \dots , \tag{7.7}$$

it turns out that:

$$\sigma_{xy}^{VC} \propto \sigma_{xy}^{noVC} + \tilde{J}^x(z_2, z_1) \left[\chi \omega \chi + \chi \omega \chi \omega \chi + \dots \right] \tilde{j}^y(z_2, z_1) . \tag{7.8}$$

Therefore, the subtraction of σ_{xy}^{noVC} from σ_{xy}^{VC} gives the contribution of vertex diagrams i.e. the extrinsic contribution to the SHE.

The next step in analyzing contributions to the SHE is to decompose the extrinsic SHE into skew scattering and side-jump scattering. For this decomposition one can use the following equation which has been taken over from the usual decomposition of the AHE [162]:

$$\sigma_{xy} = \sigma_{xy}^{skew} + \sigma_{xy}^{sj} + \sigma_{xy}^{intr} , \tag{7.9}$$

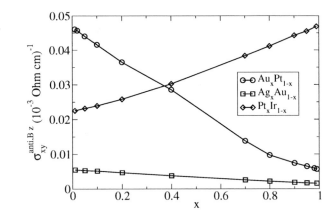

Figure 7.8: The contribution of $\sigma_{xy}^{\text{anti},B\,z}$ from Eq. (4.103) to the SHE for the transition metal alloys $\text{Au}_x\text{Pt}_{1-x}$, $\text{Ag}_x\text{Au}_{1-x}$ and $\text{Pt}_x\text{Ir}_{1-x}$.

where $\sigma_{xy}^{\text{skew}}$ corresponds to skew scattering, σ_{xy}^{sj} corresponds to side-jump scattering and $\sigma_{xy}^{\text{intr}}$ gives the intrinsic contribution (σ_{xy} has no superscript z because this relation holds also for the AHE).

A common way to display the AHE is to plot σ_{xy} versus σ_{xx} (see Fig. 7.9). Such plots can be divided into certain regions in which $\sigma_{xy} \propto \sigma_{xx}^1$ (for $\sigma_{xx} \gtrsim 10^6\,(\Omega\text{cm})^{-1}$), $\sigma_{xy} \propto \sigma_{xx}^0$ (for $\sigma_{xx} \approx 10^4 - 10^6\,(\Omega\text{cm})^{-1}$) and $\sigma_{xy} \propto \sigma_{xx}^{1.6}$ (for $\sigma_{xx} < 10^4\,(\Omega\text{cm})^{-1}$) [171]. Onoda et al. [170] theoretically showed that skew-scattering is the dominant mechanism in the superclean case ($\sigma_{xx} \gtrsim 10^6\,(\Omega\text{cm})^{-1}$) which leads to $\sigma_{xy}^{\text{skew}} = \sigma_{xx}S$ where S is the so-called skewness factor. Due to the fact that σ_{xy}^{sj} and $\sigma_{xy}^{\text{intr}}$ do not depend on impurity concentrations [153] the following relation should be valid in the superclean case:

$$\sigma_{xy} = \sigma_{xx}S + \underbrace{\sigma_{xy}^{\text{sj}} + \sigma_{xy}^{\text{intr}}}_{=\text{const}} \ . \tag{7.10}$$

Fig. 7.10 displays for $\text{Au}_x\text{Pt}_{1-x}$ the SHC versus the longitudinal conductivity σ_{xx}. It turns out that for Au with Pt concentration $\leq 10\%$ and Pt with Au concentration $\leq 10\%$ a linear behavior is obtained. Analyzing σ_{xy}^z in the spirit of Eq. (7.10) the blue and orange lines are linear fits of the data points which belong to impurity concentrations $\leq 10\%$. The intersection of the blue and orange line with the y-axis gives the side-jump and intrinsic contribution to the SHC.

The intrinsic SHE of Pt and Au have been calculated by performing cal-

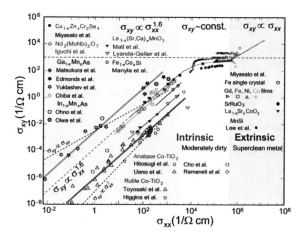

Figure 7.9: Experimental anomalous Hall conductivity vs. longitudinal conductivity for several materials. The solid magenta line shows theoretical results from Onoda et al. [171]. The figure is taken from Ref. [171].

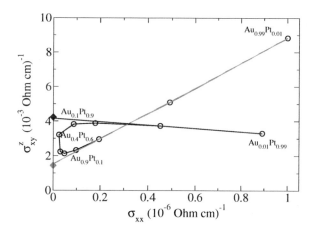

Figure 7.10: The spin Hall conductivity versus σ_{xx} for Au_xPt_{1-x} (black line/circles). The blue and orange lines/diamonds are explained in the text.

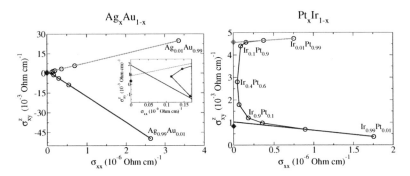

Figure 7.11: The spin Hall conductivity versus σ_{xx} (black lines/circles) for Ag_xAu_{1-x} (left panel) and Pt_xIr_{1-x} (right panel). The blue and orange lines/diamonds are explained in the text. The inset of the left panel shows a magnification of the low longitudinal conductivity regime.

culations without vertex corrections (see below). The results are represented by the blue and orange diamonds. The side-jump conductivity σ_{xy}^{sj} can be calculated as the difference between the intersections of the fitted lines with the y-axis and the corresponding diamonds. The diamonds are very close to the corresponding intersections of the blue and orange line with the y-axis which shows that side-jump scattering must be very small. A similar analysis is displayed in Fig. 7.11 for Pt_xIr_{1-x} and Ag_xAu_{1-x}. These alloy systems exhibit also a clear linear behavior in the diluted limit which shows that the relation represented by Eq. (7.10) is justified. Once again, the intersection of the linear fits with the y-axis are close to the intrinsic SHC indicating that side-jump contributions are small. However, for Ag rich Ag_xAu_{1-x} this is not the case. As demonstrated by the inset of the left panel from Fig. 7.11 there is no matching of the blue line with the blue diamond. This demonstrates that for this system an important contribution of side-jump scattering (compared to the intrinsic contribution) is present. Nevertheless, this contribution is small compared to the very big skew scattering contribution.

The intrinsic SHC can be derived from calculations without vertex corrections extrapolated down to pure metals. Fig. 7.7 shows that calculations which neglect vertex corrections match when approaching pure metals e.g. $Pt_{0.99}Ir_{0.01}$ gives nearly the same result as $Pt_{0.99}Au_{0.01}$ which indicates that the intrinsic contribution of Pt is observed. In addition, Fig. 7.7 displays results from other theoretical investigations concerning the intrinsic SHC. The quantitative agreement with the *ab initio* investigations of Guo et al. [166, 168] (these data have been multiplied by a factor of two in order to be consistent in units) and Yao and Fang [165] is good. However, deviations from the results of Tanaka et al. [96] are obtained. This can be presumably related to the fact that these results based on tight-binding calculations and therefore are less reliable as the *ab initio* results.

Using the diagrammatic representation of the Kubo-Středa equation the diagram which corresponds to the intrinsic SHE can be identified as the diagram displayed in Fig. 7.6. This diagram is approximated in the present work by the bubble diagram shown in Fig. 4.2 in which CPA averaged Green's functions are employed. However, this approximation should be well justified in the diluted case. The matching of the calculations without vertex corrections when approaching the pure metals indicates that the contributions of diagrams as shown in Fig. 7.5 are small.

As mentioned above Onoda et al. [170] investigated the crossover between an extrinsic and intrinsic dominated regime (see Fig. 7.9). They showed that the extrinsic contribution rapidly decreases as function of the inverse lifetime τ of the electronic states and drops below the intrinsic contribution. A similar analysis is shown in Fig. 7.12. The displayed extrinsic contribution is calculated via the difference between calculations including vertex corrections and calculations without vertex corrections whereas $\sigma_{xy}^{\text{intr}} = \sigma_{xy}^{\text{noVc}}$. Obviously, $\sigma_{xy}^{\text{extr}}$ decreases rapidly with decreasing σ_{xx} ($\sigma_{xx} \propto \tau$) and crosses $\sigma_{xy}^{\text{intr}}$. Due to the fact that Fig. 7.12 considers only impurity concentrations up to 30% the lifetime decrease in Ag(Au) is not enough to lead to the discussed crossover. However, the other investigated systems show clearly the crossover of the extrinsic and the intrinsic contribution which is responsible for the transition from the extrinsic to the intrinsic regime (see Fig. 7.9) with $\sigma_{xy} \propto \sigma_{xx}^0$. In addition, this figure shows that $\sigma_{xy}^{\text{intr}}$ is nearly independent of σ_{xx}.

Another interesting property of the intrinsic SHC is the sign change which appears for early 4d and 5d transition metals. Kontani et al. [172] theoretically showed that early 4d and 5d transition metals have a negative intrinsic SHC whereas the late ones exhibit a positive SHC. This is in agreement with experimental data which show for Pt [156] and Pd [173] a positive and for Nb, Mo and Ta [173] a negative intrinsic SHC (the negative SHC of Nb is also mentioned in Ref. [172] as unpublished data from Y. Otani *et al.*). Fig. 7.13 shows the SHC for two bcc systems which belong to the early 4d and 5d transition metals: $\text{Nb}_x\text{Mo}_{1-x}$ and $\text{W}_x\text{Ta}_{1-x}$, respectively. In addition, Fig. 7.14 displays the contribution of the term $\sigma_{xy}^{\text{anti},B\,z}$ from Eq. (4.103) and Fig. 7.15 shows an analysis similar to Fig. 7.10. As in the case of late transition metals the contribution of the term $\sigma_{xy}^{\text{anti},B\,z}$ is very small. In agreement with the above mentioned theoretical and experimental observations the intrin-

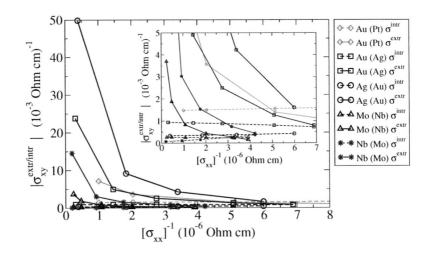

Figure 7.12: The extrinsic/intrinsic contribution to the SHE versus the inverse of the longitudinal conductivity for several alloy systems. The atom type in brackets denotes the impurity type (impurity concentrations up to 30% are considered). The solid lines correspond to $\sigma_{xy}^{\text{extr}}$ and the dashed lines correspond to $\sigma_{xy}^{\text{intr}}$.

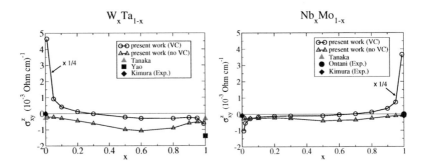

Figure 7.13: The spin Hall conductivity for $W_x Ta_{1-x}$ and $Nb_x Mo_{1-x}$. In addition, theoretical data [96, 165] as well as experimental data are shown [172, 173].

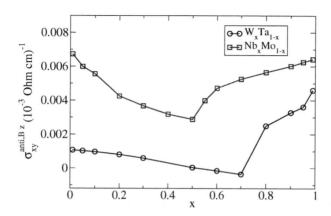

Figure 7.14: The contribution of $\sigma_{xy}^{\text{anti},B\,z}$ from Eq. (4.103) to the SHC for the transition metal alloys $W_x Ta_{1-x}$ and $Nb_x Mo_{1-x}$.

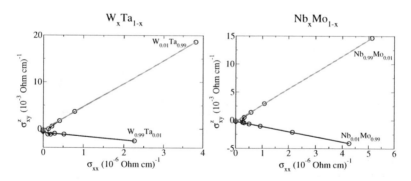

Figure 7.15: The spin Hall conductivity versus σ_{xx} (black lines/circles) for $W_x Ta_{1-x}$ (left panel) and $Nb_x Mo_{1-x}$ (right panel). The blue and orange lines/diamonds are explained in the text.

alloy system	intrinsic $\sigma_{xy}^{\text{intr}}$		skew scattering $\sigma_{xy}^{\text{skew}}$		side-jump σ_{xy}^{sj}
$\text{Au}_x\text{Pt}_{1-x}$	Au	1.44	$\text{Au}_{0.99}\text{Pt}_{0.01}$	7.31	0.08
	Pt	4.23	$\text{Au}_{0.01}\text{Pt}_{0.99}$	-0.74	-0.17
$\text{Ag}_x\text{Au}_{1-x}$	Ag	0.32	$\text{Ag}_{0.99}\text{Au}_{0.01}$	-51.52	1.58
	Au	0.95	$\text{Ag}_{0.01}\text{Au}_{0.99}$	24.00	-0.12
$\text{Pt}_x\text{Ir}_{1-x}$	Pt	4.58	$\text{Pt}_{0.99}\text{Ir}_{0.01}$	0.17	-0.02
	Ir	0.84	$\text{Pt}_{0.01}\text{Ir}_{0.99}$	-0.65	0.18
$\text{W}_x\text{Ta}_{1-x}$	W	-0.64	$\text{W}_{0.99}\text{Ta}_{0.01}$	-1.90	-0.13
	Ta	-0.19	$\text{W}_{0.01}\text{Ta}_{0.99}$	18.80	-0.08
$\text{Nb}_x\text{Mo}_{1-x}$	Nb	-0.07	$\text{Nb}_{0.99}\text{Mo}_{0.01}$	14.73	-0.18
	Mo	-0.28	$\text{Nb}_{0.01}\text{Mo}_{0.99}$	-3.80	0.05

Table 7.1: SHE for several $4d$ and $5d$ transition metals decomposed into intrinsic, skew scattering and side-jump contribution (all results in $10^3(\Omega\,\text{cm})^{-1}$).

sic SHC which is derived from an extrapolation of the calculations without vertex corrections to the pure metals exhibits a negative sign and is much smaller as e.g. for Pt. The analysis of the SHC results for $\text{W}_x\text{Ta}_{1-x}$ and $\text{Nb}_x\text{Mo}_{1-x}$ which is shown in Fig. 7.15 displays a clear linear dependence of σ_{xy}^z from σ_{xx} what indicates that again Eq. (7.10) is fulfilled.

Tab. 7.1 summarizes the observations of the present section concerning the SHC. $\sigma_{xy}^{\text{intr}}$ is derived from the extrapolation of $\sigma_{xy}^{\text{noVC}}$ to the pure metals whereas $\sigma_{xy}^{\text{skew}}$ and σ_{xy}^{sj} are calculated from analysis shown in Figs. 7.10, 7.11 and 7.15. It is important to note that the meaning of the numbers for σ_{xy}^{sj} is limited. Due to the fact that σ_{xy}^{sj} is very small (apart from Ag with Au impurities) the indirect calculation of the side-jump contribution via a fitting procedure leads to a dependency of the results from e.g. the chosen data points which are considered within the fitting procedure. Therefore, the conclusion is that side-jump scattering is small for all systems compared to skew scattering and for nearly all systems small compared to the intrinsic contribution. Depending on the system σ_{xy}^{sj} shows a tendency to a positive or negative sign independent of the sign of $\sigma_{xy}^{\text{skew}}$. There seems to be no correlation of the sign of σ_{xy}^{sj} to that of $\sigma_{xy}^{\text{intr}}$ for the systems studied here.

In order to proof that the analysis used for the decomposition of σ_{xy}^z is justified one can compare $\sigma_{xy}^{\text{skew}}$ which is calculated via $\sigma_{xy}^{\text{skew}} = \sigma_{xx}S$ (S is derived from the slope of the linear fits) with a direct calculation of the skew scattering contribution from the Boltzmann equation (see appendix D). Fig. 7.16 displays a comparison of the results from the present work with calculations

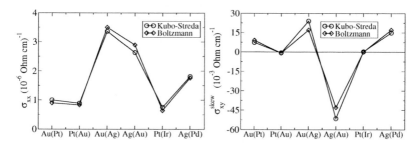

Figure 7.16: The longitudinal conductivity σ_{xx} (left panel) and $\sigma_{xy}^{\text{skew}}$ (right panel) calculated with the Kubo-Středa and the Boltzmann formalism [174] for impurity concentrations of 1%. The atom type within the brackets denotes the impurity type.

using the Boltzmann formalism [174]. This figure demonstrates that a very good quantitative agreement is achieved for σ_{xx} as well as $\sigma_{xy}^{\text{skew}}$. This clearly shows that the decomposition procedure of the SHC in skew scattering and side-jump scattering as performed in the present work is well justified.

The quantitative agreement of $\sigma_{xy}^{\text{intr}}$ with experimental data for the intrinsic SHC is good for early transition metals like e.g. Nb whereas for Pt a significant deviation is observed in the present work as well as in Ref. [166]. Guo et al. [166] theoretically investigated the influence of temperature on the SHC. They showed that the intrinsic SHC in Pt rapidly decreases with increasing temperature and obtained a very good agreement of the calculated intrinsic SHC at room temperature compared to experiment. At first sight the observation of a decreasing SHC with increasing temperature is in variance to measurements of Vila et al. [156] which observed a nearly temperature independent SHC in the range from $\approx$ 5-300 K. This contradiction can be presumably explained by the fact that even at very low temperatures the measurements show a longitudinal conductivity of $\sigma_{xx} \approx 10^5$ $(\Omega\,\text{cm})^{-1}$ which is much lower compared to the superclean regime assumed in the work of Guo et al. [166]. Therefore a direct comparison is hindered.

The observation that small impurity concentrations can drastically increase the SHC is in agreement with recent experimental measurements. Seki et al. [157] measured a gigantic SHC in Au ($\sigma_{xy}^{z} \propto 10^5\,(\Omega\text{cm})^{-1}$). They conclude that the large SHC is based on extrinsic contributions (skew scattering) from impurities in Au. Similar observations are made by Koong et al. [175] which measured a huge SHC in Pt and attributed this the skew scattering mecha-

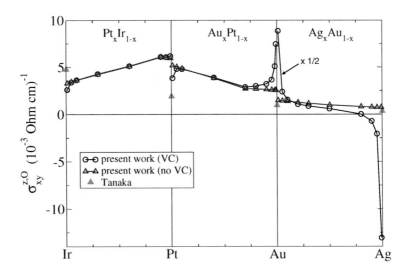

Figure 7.17: OrbHC for three different alloy systems: Pt_xIr_{1-x}, Au_xPt_{1-x} and Ag_xAu_{1-x}. The circles (black line) correspond to calculations including vertex corrections (the VC results of Ag_xAu_{1-x} are multiplied by 0.5) and the triangles (blue line) represent calculations without vertex corrections. In addition, theoretical data from Tanaka et al. [96] for the intrinsic OrbHC are shown.

nism.

The family of Hall effects contains as further related effect the so-called orbital Hall effect (OrbHE) [96, 172, 176]. This effect can be calculated via the Kubo-Středa equation:

$$\sigma_{xy}^{z,O} = \frac{\hbar}{2\pi V}\text{Tr}\left\langle \hat{J}_x^{z,O} G^+ \hat{j}_y G^- \right\rangle_c + \frac{e}{4\pi iV}\text{Tr}\left\langle (G^+ - G^-)(\hat{r}_x \hat{J}_y^{z,O} - \hat{r}_y \hat{J}_x^{z,O}) \right\rangle_c,$$

$$(7.11)$$

with $\hat{J}_x^{z,O} = e\,\hat{l}_z\,\hat{j}_x$ ($\hat{l}_z$ is the z-component of the angular momentum operator, $e = |e|$). The OrbHE corresponds to a flow of atomic orbital angular momentum in a transverse direction compared to the applied electric field. Theoretical investigations predict the intrinsic OrbHE in transition metals [96] and oxides [177] even larger than the intrinsic SHE. Figs. 7.17 and 7.18 show the orbital Hall conductivity (OrbHC) for late and early transition metals, respectively. For comparison theoretical data for the intrinsic

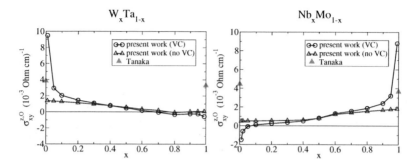

Figure 7.18: OrbHC for W_xTa_{1-x} (left panel) and Nb_xMo_{1-x} (right panel). The circles (black line) correspond to calculations including vertex corrections and the triangles (blue line) represent calculations without vertex corrections. In addition, theoretical data from Tanaka et al. [96] for the intrinsic OrbHC are shown.

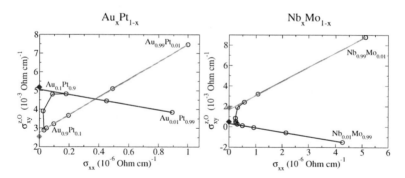

Figure 7.19: The orbital Hall conductivity versus σ_{xx} (black lines/circles) for Au_xPt_{1-x} (left panel) and Nb_xMo_{1-x} (right panel). The blue and orange lines/diamonds are explained in the text.

OrbHC from Tanaka et al. [96] are displayed. These data have been derived from tight-binding model calculations which presumably explains the quantitative deviations for some metals. However, the qualitative agreement is reasonable. Tanaka et al. [96] predicted no sign change of the intrinsic OrbHC over the whole range of $4d$ and $5d$ transition metals what is in agreement with the present work. Fig. 7.19 shows a similar analysis as displayed in e.g. Fig. 7.10 for an early and late transition metal alloy system. The OrbHC shows as the SHC a linear behavior for the diluted case allowing an analysis employing Eq. (7.10). The comparison of the intersections of the blue and orange line with the y-axis with the corresponding intrinsic OrbHC shows that the side-jump contribution is small. The dominant mechanism is skew scattering with the exception of Pt with small Au concentrations where the large intrinsic effect prevails.

7.2 Anomalous Hall Effect

In this section the AHE is investigated for three different fcc alloy systems: Fe_xPd_{1-x}, Co_xPd_{1-x} and Ni_xPd_{1-x}. These alloy systems are chosen due to the fact that experimental data over a wide concentration range are available. In addition to the anomalous Hall conductivity (AHC) the isotropic residual resistivity $\rho = \frac{1}{3}\rho_\parallel + \frac{2}{3}\rho_\perp$ and the anisotropic magnetoresistance (AMR) are shown for each alloy system. The AMR is defined as [178]:

$$\frac{\Delta\rho}{\rho} = \frac{\rho_\parallel - \rho_\perp}{\frac{1}{3}\rho_\parallel + \frac{2}{3}\rho_\perp} \, , \tag{7.12}$$

where $\rho_\parallel$ and $\rho_\perp$ are the resistivity parallel and perpendicular to the magnetization direction, respectively. Similar to the AHE, the AMR is a purely relativistic effect which is caused by spin-orbit coupling.

Fig. 7.20 shows the residual resistivity as well as the AMR for Fe_xPd_{1-x}. The comparison with experimental data shows good agreement. However, one has to take into account that for the calculations a fcc lattice has been used whereas experimental investigations indicate the possibility of a face-centered tetragonal phase for Fe concentrations between $\approx$ 37-55% (at higher Fe concentrations a bcc phase coexists with the face-centered phases) [181]. Fig. 7.21 shows the AHC as well as the contribution of the term $\sigma_{xy}^{\text{anti},B}$ from Eq. (4.103) for Fe_xPd_{1-x}. In order to compare the calculated AHC with experiment one has to use the results including vertex corrections. Obviously, a satisfying agreement with the experimental data is obtained. Even the experimental sign change is in excellent agreement with the theoretical one. In Sec. 4.6.3 it was claimed that the contribution of the term $\sigma_{xy}^{\text{anti},B}$ is

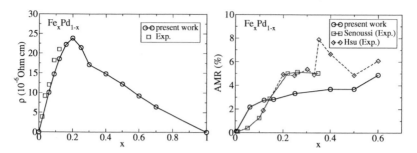

Figure 7.20: Residual resistivity (left figure) and AMR (right figure) of Fe_xPd_{1-x}. The experimental resistivity data are from Skalski et al. [179] (4.2 K) and the experimental AMR data are from Senoussi et al. [180] (1.5 K) as well as Hsu et al. [181] (4.2 K).

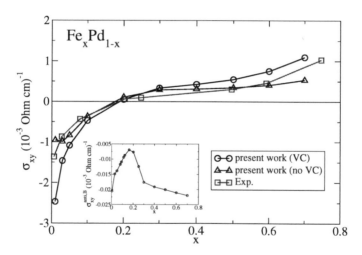

Figure 7.21: The AHC of Fe_xPd_{1-x}. The circles show results including vertex corrections and the triangles show results without vertex corrections. In addition, the inset displays the small contribution of the term $\sigma_{xy}^{anti,B}$ from Eq. (4.103). The squares show experimental data from Ref. [182] (4.2 K).

negligible compared to the term $\sigma_{xy}^{anti,A}$. This is demonstrated by the insets of Figs. 7.21, 7.23 and 7.25.

Figs. 7.22 and 7.23 show the residual resistivity, AMR and the AHE

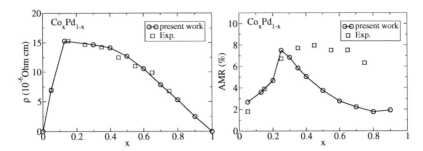

Figure 7.22: Residual resistivity (left figure) and AMR (right figure) of Co_xPd_{1-x}. The experimental resistivity data are from Jen et al. [183] (4 K) and the experimental AMR data are from Jen [184] (4 K).

for Co_xPd_{1-x}, respectively. Again, the agreement with experimental data is good. However, the AMR shows some deviations at higher Co concentrations which could be due to structural deviations from the pure fcc phase. The comparison of the AHC with experiment shows for Co concentrations $\gtrsim 8\%$ a very good agreement but for lower Co concentrations a deviation from experiment is obtained. Due to the fact that with decreasing impurity concentration the disorder induced broadening of the electronic states decreases the Brillouin zone integration (see Sec. 4.6.1) becomes numerically very demanding. Therefore, the experimental sign change of the AHC for diluted Co concentrations is difficult to reproduce. The smallest Co concentration considered is only 1% and therefore one can not preclude that calculations using lower Co concentrations show the experimental sign change (impurity concentrations down to 0.2% indicate no sign change in the present work).

For the alloy system Ni_xPd_{1-x} the residual resistivity and the AMR are shown in Fig. 7.24 and the AHC is shown in Fig. 7.25. For this system the residual resistivity shows a systematic underestimation over the whole concentration range. Usually the residual resistivity calculated within the formalism used in the present work is in good agreement with low temperature measurements (see e.g. Figs. 6.21, 7.20 and 7.22). Therefore, the pronounced deviation could presumably be related to lattice imperfections in the experimental investigated samples. Lattice imperfections of any kind lead to an increase of the measured resistivity [3]. Nevertheless, the AHC effect shows good quantitative agreement with the experimental data and the sign change is in line with the experimental observations.

If one subtracts from calculations including vertex corrections the results

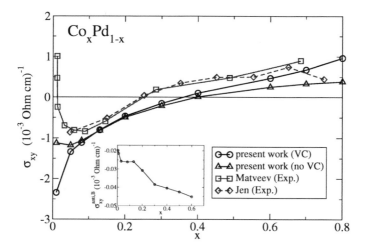

Figure 7.23: The AHC of Co_xPd_{1-x}. The circles show results including vertex corrections and the triangles show results without vertex corrections. In addition, the inset displays the small contribution of the term $\sigma_{xy}^{anti,B}$ from Eq. (4.103). The squares and diamonds show experimental data from Ref. [182] and [185] (4.2 K), respectively.

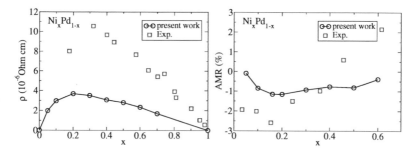

Figure 7.24: Residual resistivity (left figure) and AMR (right figure) of Ni_xPd_{1-x}. The experimental resistivity data are from Dreesen and Pugh [186] (4 K) and the experimental AMR data are from Senoussi et al. [180] (1.5 K).

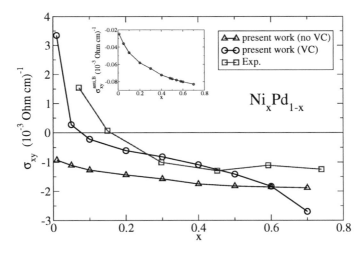

Figure 7.25: The AHC of Ni_xPd_{1-x}. The circles show results including vertex corrections and the triangles show results without vertex corrections. In addition, the inset displays the small contribution of the term $\sigma_{xy}^{anti,B}$ from Eq. (4.103). The squares show experimental data from Ref. [182] (4.2 K).

from calculations without vertex corrections one gets the purely extrinsic contribution to the AHE (see Sec. 7.1). In order to decompose the extrinsic AHC into skew and side-jump scattering one can employ Eq. (7.10) without the intrinsic contribution i.e. $\sigma_{xy}^{extr} = \sigma_{xx}S + \sigma_{xy}^{sj}$. Fig. 7.26 shows the extrinsic AHC versus the longitudinal conductivity for Pd with impurity concentrations up to 10%. Obviously, a linear behavior due to skew scattering is observed. The brown, magenta and orange lines are linear fits of the corresponding calculated data points. The intersections of the fitted lines with the y-axis give the side-jump scattering contribution which is very small compared to skew scattering for all three alloy systems. This analysis is very similar to the analysis used for the spin Hall effect (see e.g. Fig. 7.10). However, Fig. 7.26 shows only the extrinsic contribution to the AHC whereas Fig. 7.10 shows the sum of extrinsic and intrinsic SHC.

Due to the fact that non-magnetic metals exhibit no intrinsic AHE the calculations which neglect vertex corrections should go down to zero when approaching Pd. At first sight this is in contradiction with the results shown in Figs. 7.21, 7.23 and 7.25 where the calculations without vertex corrections still give $\approx -10^3$ Ωcm for 1% impurity concentration. Numerically very demanding calculations (up to $\approx 10^9$ k-points within the first Brillouin zone

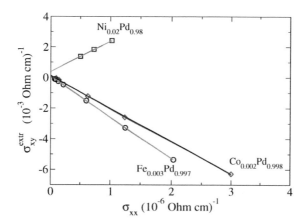

Figure 7.26: The extrinsic AHC versus σ_{xx} for Ni_xPd_{1-x}, Co_xPd_{1-x} and Fe_xPd_{1-x}. The brown, magenta and orange lines are explained in the text.

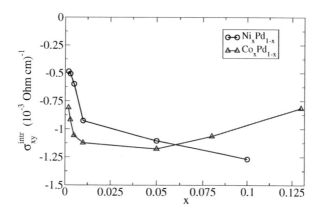

Figure 7.27: The AHC of Ni_xPd_{1-x} and Co_xPd_{1-x}calculated without vertex corrections for diluted impurity concentrations.

have been used) for slightly lower impurity concentrations indicate that the calculations without vertex corrections approaches zero with decreasing impurity concentration. This behavior is shown in Fig. 7.27 for Ni_xPd_{1-x} and Co_xPd_{1-x}. Due to computational limitations concerning the k-space integration impurity concentrations $< 0.2\%$ have not been investigated within the present work.

In summary, this section demonstrates that the method used in the present work is suitable for investigations of the AHE. The quantitative agreement with experimental data is good. Only for very diluted impurity concentrations deviations from experimental data are observed. It turns out that for low impurity concentrations skew scattering is the dominant mechanism.

Chapter 8

Summary

The major aim of the present work was to investigate theoretically the symmetric as well as anti-symmetric part of the electric conductivity tensor for alloys on a relativistic *ab initio* level. This has been achieved using the spin-polarized fully relativistic Korringa-Kohn-Rostoker Green's function method (SPR-KKR-GF) in conjunction with the linear response Kubo-Středa formalism. For the investigation of the SHE a fully relativistic spin-current density operator derived from the Bargmann-Wigner four-vector polarization operator $\mathcal{T}$ has been used. In order to investigate spin dependent phenomena spin projection operators derived from $\mathcal{T}$ have been employed. The simulation of the electronic structure of random disordered alloys has been performed using the coherent potential approximation (CPA). In addition, the non-local coherent potential approximation (NLCPA) has been employed to include short ranged correlations in the atomic lattice site occupation.

In particular the theoretical investigation of the anti-symmetric part of the conductivity tensor, i.e. the Hall conductivity coefficients, is very challenging. In order to calculate the intrinsic SHE/AHE one needs a transport theory which gives access to the anti-symmetric part of the conductivity tensor in combination with a proper relativistic treatment (inclusion of spin-orbit coupling) of the electronic structure. In addition, for the calculation of the SHE a spin decomposition scheme has to be applied. Recently, it turned out that the SHE/AHE is very sensitive to impurities which give rise to the extrinsic SHE/AHE. To investigate the extrinsic SHE/AHE the theoretical framework has to be extended to a reliable treatment of disorder. The combination of all the mentioned requirements makes the *ab initio* calculation of the SHE/AHE very challenging. Therefore, only few theoretical investigations are available which consider the intrinsic AHE and SHE on an *ab initio* level or the intrinsic/extrinsic contribution via model Hamiltonians. The method used in the present work allows the first simultaneous investigation of the extrinsic as well as intrinsic contribution to the SHE/AHE on an *ab initio* level. This has been demonstrated for several non-magnetic 4d- and 5d-transition metal alloy systems (SHE) and for magnetic Pd based alloys with 3d-transition metals (AHE). Due to the fact that intrinsic as well as extrinsic contributions are investigated on equal footings it was for the first time possible to decompose the SHE/AHE in a reliable way into skew and side-jump scattering as well as intrinsic contributions.

One of the central observations is that for all investigated alloy systems skew scattering is the dominant mechanism within the investigated super-clean regime. However, for impurity concentrations of 1% the intrinsic contribution can exceed the skew scattering contribution. Such a behavior was observed for the SHE of $Pt_x Ir_{1-x}$ where the vertex corrections (VC) are small for impurity concentrations $\geq 1\%$ which indicates that the skew scattering dominated regime for this system emerges for impurity concentrations below 1%. It turns out that the side-jump contribution is negligible for most of the investigated systems within the superclean regime which is in agreement with earlier theories. In order to show the reliability of the decomposition method used within the present work to extract skew and side-jump scattering the SHE results for skew scattering are compared with calculations using the Boltzmann formalism which show good quantitative agreement.

The present work clarified the ongoing debate about the importance of VC. It was demonstrated that the inclusion of VC which give rise to the skew scattering contribution is essential for the calculation of the SHE/AHE within the superclean regime. The observation that the importance of VC increases with decreasing impurity concentration is in contrast to the observations within standard transport calculations like the residual resistivity where VC become negligible in the very diluted case. In addition, it was shown that with going from the dilute to concentrated concentration regime the importance of VC for the SHE/AHE drop down which lead to a switching from an extrinsic to an "intrinsic" regime dominated by intrinsic instead of extrinsic effects. Therefore, the present work clearly demonstrates that experimental investigations of the intrinsic SHE using metallic samples of very high purity are extremely sensitive to impurity atoms which can influence the measurements in a very pronounced way. This leads to the counterintuitive observation that such measurements should be performed using samples with well-defined impurity concentrations i.e. depending on the impurity type one has to go beyond a certain impurity concentration in order to reduce the diverging skew scattering contribution with decreasing impurity concentration. The comparison of the calculations with experimental data show especially in the case of the AHE an excellent quantitative agreement. All experimentally observed sign changes for concentrated alloys could be reproduced. In particular, the calculated sign change of the intrinsic SHE from early to late transition metals is in line with experimental data.

In order to decompose the conductivity tensor elements into "spin-up" and "spin-down" contributions a spin decomposition scheme had to be used. The present work derives suitable spin projection operators from the relativistic Bargmann-Wigner four-vector polarization operator which can be considered as a generalized spin operator. These spin projection operators have been applied to the symmetric part of the conductivity tensor. The results are compared to an approximate spin decomposition scheme which shows a very good agreement for alloy systems with negligible spin-flip contributions (low spin-orbit coupling). In addition, it has been demonstrated that for systems with strong spin-orbit coupling the approximate scheme fails whereas the spin projection operators give reliable results.

Several alloys show instead of random disorder short ranged correlations in the lattice site occupation. This can drastically influence the transport properties of the alloys. In order to investigate the influence of such short ranged correlations one has to go beyond the CPA. The NLCPA is a cluster generalization of the CPA which allows to simulate short ranged ordering effects. Usually, the residual resistivity decreases with increasing ordering. This is demonstrated for Cu_xZn_{1-x} within the present work. However, there is a significant set of alloys which show a completely opposite behavior so that their resistivities actually increase when ordering is increased. These materials belong to the class of so-called K-state alloys. Such alloys are typically rich in late transition metals such as Ni, Pd or Pt and alloyed with a mid-row element such as Cr, Mo or W. The K-effect has been investigated for $Ni_{0.8}Mo_{0.2}$, $Ni_{0.8}Cr_{0.2}$ $Pd_{0.8}W_{0.2}$ and Ag_xPd_{1-x}. The NLCPA results confirm a decrease in resistivity with decreasing ordering. For Ag_xPd_{1-x} the K-effect is only observed for the Pd rich side. The present work proposes an explanation of the K-effect which is caused by characteristic features of the electronic states at the Fermi energy. These features could be observed and identified for all investigated K-state alloys.

Chapter 8.   Summary

Appendix A

Wave function of a Scattered Electron for $r \geq r_{\mathrm{mt}}$

In order to calculate the wave function of a scattered electron outside the potential region ($V(\mathbf{r}) = 0$) one can use Eq. (3.27) in conjunction with Eqs. (3.9), (3.15) and (3.23). This leads to the following expression

$$
\psi(\mathbf{r}) = \sum_{\kappa,\mu} \begin{pmatrix} j_l(pr)\chi_\kappa^\mu(\hat{\mathbf{r}}) \\ \omega\, j_{\bar{l}}(pr)\chi_{-\kappa}^\mu(\hat{\mathbf{r}}) \end{pmatrix}
$$
$$
- \sum_{\substack{\kappa,\mu \\ \kappa',\mu'}} \int_0^{r_{\mathrm{mt}}} \int dr' d\hat{\mathbf{r}}'\, r'^2\, i\, p\, \eta \begin{pmatrix} h_l^+(pr)\chi_\kappa^\mu(\hat{\mathbf{r}}) \\ \omega\, h_{\bar{l}}^+(pr)\chi_{-\kappa}^\mu(\hat{\mathbf{r}}) \end{pmatrix}
$$
$$
\begin{pmatrix} j_l(pr')\chi_\kappa^\mu(\hat{\mathbf{r}}') \\ \omega\, j_{\bar{l}}(pr')\chi_{-\kappa}^\mu(\hat{\mathbf{r}}') \end{pmatrix}^\times V(\mathbf{r}')\mathbb{1}_4 \begin{pmatrix} g_{\kappa'}(r')\chi_{\kappa'}^{\mu'}(\hat{\mathbf{r}}') \\ if_{\kappa'}(r')\chi_{-\kappa'}^{\mu'}(\hat{\mathbf{r}}') \end{pmatrix} \tag{A.1}
$$

for $r \geq r_{\mathrm{mt}}$ with $\eta = \frac{E+mc^2}{c^2}$ and $\omega = \frac{ipcS_\kappa}{E+mc^2}$. The last equation can be written as

$$
\psi(\mathbf{r}) = \sum_{\kappa,\mu} \begin{pmatrix} j_l(pr)\chi_\kappa^\mu(\hat{\mathbf{r}}) \\ \omega\, j_{\bar{l}}(pr)\chi_{-\kappa}^\mu(\hat{\mathbf{r}}) \end{pmatrix} - \sum_{\kappa,\mu} i\, p\, \eta \begin{pmatrix} h_l^+(pr)\chi_\kappa^\mu(\hat{\mathbf{r}}) \\ \omega\, h_{\bar{l}}^+(pr)\chi_{-\kappa}^\mu(\hat{\mathbf{r}}) \end{pmatrix} t_{\kappa,\mu} \tag{A.2}
$$

with

$$
t_{\kappa,\mu} = \sum_{\kappa',\mu'} \int_0^{r_{\mathrm{mt}}} \int dr\, d\hat{\mathbf{r}}\, r^2 \begin{pmatrix} j_l(pr)\chi_\kappa^\mu(\hat{\mathbf{r}}) \\ \omega\, j_{\bar{l}}(pr)\chi_{-\kappa}^\mu(\hat{\mathbf{r}}) \end{pmatrix}^\times V(\mathbf{r})\mathbb{1}_4 \begin{pmatrix} g_{\kappa'}(r)\chi_{\kappa'}^{\mu'}(\hat{\mathbf{r}}) \\ if_{\kappa'}(r)\chi_{-\kappa'}^{\mu'}(\hat{\mathbf{r}}) \end{pmatrix} \tag{A.3}
$$

$$
= \int_0^{r_{\mathrm{mt}}} dr\, r^2 \begin{pmatrix} j_l(pr) \\ \omega\, j_{\bar{l}}(pr) \end{pmatrix}^\times V(r)\mathbb{1}_2 \begin{pmatrix} g_{\kappa'}(r) \\ if_{\kappa'}(r) \end{pmatrix}. \tag{A.4}
$$

For the derivation of Eq. (A.4) the identity

$$\int d\hat{\mathbf{r}} \, \chi_\kappa^\mu(\hat{\mathbf{r}})^\dagger \chi_{\kappa'}^{\mu'}(\hat{\mathbf{r}}) = \delta_{\kappa,\kappa'} \delta_{\mu,\mu'} \tag{A.5}$$

has been used in combination with the assumption that the potential $V(\mathbf{r})$ contains no terms that couple to the spin-angular functions (this would be the case if the potential includes a magnetic field). Finally, the wave function of the scattered electron becomes

$$\psi(\mathbf{r}) = \sum_{\kappa,\mu} \begin{pmatrix} j_l(pr)\chi_\kappa^\mu(\hat{\mathbf{r}}) \\ \omega \, j_{\bar{l}}(pr)\chi_{-\kappa}^{\mu'}(\hat{\mathbf{r}}) \end{pmatrix} - \sum_{\kappa,\mu} i \, p \, \eta \begin{pmatrix} (j_l(pr) + i n_l(pr))\chi_\kappa^\mu(\hat{\mathbf{r}}) \\ \omega \, (j_{\bar{l}}(pr) + i n_{\bar{l}}(pr))\chi_{-\kappa}^\mu(\hat{\mathbf{r}}) \end{pmatrix} t_\kappa \tag{A.6}$$

where Eq. (3.13) has been used. If one compares Eq. (A.6) with Eq. (3.14) it turns out that the coefficients of Eq. (3.14) are directly related to t_κ via

$$\cos \delta_\kappa = 1 - i \, p \, \eta \, t_\kappa \qquad \sin \delta_\kappa = -p \, \eta \, t_\kappa \, . \tag{A.7}$$

This observation shows that it is sufficient to calculate the phase shifts δ_κ to determine t_κ.

A more general method for the determination of t_κ was derived by Ebert and Györffy [187]. This method uses a Wronskian relation for the calculation of t_κ and is also applicable in the case of non-spherically-symmetric potentials.

Appendix B

Cluster Probabilities

In order to apply the NLCPA a crucial step is the proper definition of cluster probabilities P_γ. In the present work only the smallest fcc (four atoms) and bcc (two atoms) clusters are used. This lead for a binary bcc (fcc) alloy to 4 (16) cluster configurations. The corresponding cluster probabilities P_γ are shown in Tab. B.1 and Tab. B.2.

cluster configuration	disorder $P_\gamma =$	clustering $P_\gamma =$	ordering
Cu,Cu	x^2	x	$P_\gamma = 2x - 1$ if $x \geq \frac{1}{2}$; $P_\gamma = 0$ if $x < \frac{1}{2}$
Cu,Zn	$x(1-x)$	0	$P_\gamma = 1 - x$ if $x \geq \frac{1}{2}$; $P_\gamma = x$ if $x < \frac{1}{2}$
Zn,Cu	$x(1-x)$	0	$P_\gamma = 1 - x$ if $x \geq \frac{1}{2}$; $P_\gamma = x$ if $x < \frac{1}{2}$
Zn,Zn	$(1-x)^2$	$1-x$	$P_\gamma = 0$ if $x \geq \frac{1}{2}$; $P_\gamma = 1 - 2x$ if $x < \frac{1}{2}$

Table B.1: Cluster probabilities for the NLCPA bcc cluster (shown for Cu_xZn_{1-x}).

cluster configuration	disorder $P_\gamma =$	clustering $P_\gamma =$	ordering
Ni,Ni,Ni,Ni	x^4	x	$P_\gamma = 0$ if $x < 0.75$; $P_\gamma = 4x - 3$ if $x \geq 0.75$
Ni,Ni,Ni,Mo	$x^3(1-x)$	0	$P_\gamma = 0$ if $x < 0.5$; $P_\gamma = x - 0.5$ if $0.5 \leq x < 0.75$; $P_\gamma = 1 - x$ if $x \geq 0.75$
Ni,Ni,Mo,Ni	$x^3(1-x)$	0	$P_\gamma = 0$ if $x < 0.5$; $P_\gamma = x - 0.5$ if $0.5 \leq x < 0.75$; $P_\gamma = 1 - x$ if $x \geq 0.75$
Ni,Mo,Ni,Ni	$x^3(1-x)$	0	$P_\gamma = 0$ if $x < 0.5$; $P_\gamma = x - 0.5$ if $0.5 \leq x < 0.75$; $P_\gamma = 1 - x$ if $x \geq 0.75$
Mo,Ni,Ni,Ni	$x^3(1-x)$	0	$P_\gamma = 0$ if $x < 0.5$; $P_\gamma = x - 0.5$ if $0.5 \leq x < 0.75$; $P_\gamma = 1 - x$ if $x \geq 0.75$
Ni,Ni,Mo,Mo	$x^2(1-x)^2$	0	$P_\gamma = 0$ if $x < 0.25$; $P_\gamma = \frac{2}{3}x - \frac{1}{6}$ if $0.25 \leq x < 0.5$; $P_\gamma = -\frac{2}{3}x + \frac{1}{2}$ if $x \geq 0.5$
Ni,Mo,Ni,Mo	$x^2(1-x)^2$	0	$P_\gamma = 0$ if $x < 0.25$; $P_\gamma = \frac{2}{3}x - \frac{1}{6}$ if $0.25 \leq x < 0.5$; $P_\gamma = -\frac{2}{3}x + \frac{1}{2}$ if $x \geq 0.5$
Mo,Ni,Mo,Ni	$x^2(1-x)^2$	0	$P_\gamma = 0$ if $x < 0.25$; $P_\gamma = \frac{2}{3}x - \frac{1}{6}$ if $0.25 \leq x < 0.5$; $P_\gamma = -\frac{2}{3}x + \frac{1}{2}$ if $x \geq 0.5$
Mo,Ni,Ni,Mo	$x^2(1-x)^2$	0	$P_\gamma = 0$ if $x < 0.25$; $P_\gamma = \frac{2}{3}x - \frac{1}{6}$ if $0.25 \leq x < 0.5$; $P_\gamma = -\frac{2}{3}x + \frac{1}{2}$ if $x \geq 0.5$
Ni,Mo,Mo,Ni	$x^2(1-x)^2$	0	$P_\gamma = 0$ if $x < 0.25$; $P_\gamma = \frac{2}{3}x - \frac{1}{6}$ if $0.25 \leq x < 0.5$; $P_\gamma = -\frac{2}{3}x + \frac{1}{2}$ if $x \geq 0.5$
Mo,Mo,Ni,Ni	$x^2(1-x)^2$	0	$P_\gamma = 0$ if $x < 0.25$; $P_\gamma = \frac{2}{3}x - \frac{1}{6}$ if $0.25 \leq x < 0.5$; $P_\gamma = -\frac{2}{3}x + \frac{1}{2}$ if $x \geq 0.5$
Mo,Mo,Mo,Ni	$x(1-x)^3$	0	$P_\gamma = x$ if $x < 0.25$; $P_\gamma = 0.5 - x$ if $0.25 \leq x < 0.5$; $P_\gamma = 0$ if $x \geq 0.5$
Mo,Mo,Ni,Mo	$x(1-x)^3$	0	$P_\gamma = x$ if $x < 0.25$; $P_\gamma = 0.5 - x$ if $0.25 \leq x < 0.5$; $P_\gamma = 0$ if $x \geq 0.5$
Mo,Ni,Mo,Mo	$x(1-x)^3$	0	$P_\gamma = x$ if $x < 0.25$; $P_\gamma = 0.5 - x$ if $0.25 \leq x < 0.5$; $P_\gamma = 0$ if $x \geq 0.5$
Ni,Mo,Mo,Mo	$x(1-x)^3$	0	$P_\gamma = x$ if $x < 0.25$; $P_\gamma = 0.5 - x$ if $0.25 \leq x < 0.5$; $P_\gamma = 0$ if $x \geq 0.5$
Mo,Mo,Mo,Mo	$(1-x)^4$	$1 - x$	$P_\gamma = 1 - 4x$ if $x < 0.25$; $P_\gamma = 0$ if $x \geq 0.25$

Table B.2: Cluster probabilities for the NLCPA fcc cluster (shown for Ni_xMo_{1-x}).

Appendix C

Weak-Disorder Limit

The diagrammatic representation of the conductivity in Sec. 4.5 considers only terms which consists of a combination of retarded and advanced Green's functions. The reason for this procedure is that in the weak-disorder limit the contributions of terms which consists of a combination of two retarded or two advanced Green's functions give a negligible contribution to the conductivity. In order to proof this statement one can start from Eq. (4.62) with two retarded Green's functions. In case of negligible vertex corrections (Refs. [63, 88] discuss also the more general case) this equation becomes proportional to $\langle G^+ \rangle \langle G^+ \rangle$. With the help of Eq. (3.65) one can write

$$
\langle G^+ \rangle \langle G^+ \rangle = \frac{1}{(\underbrace{E - H - \Re\Sigma}_{=x} \underbrace{-i \Im\Sigma}_{=-\Delta})^2} = \frac{1}{(x + i\Delta)^2} \times \frac{(x - i\Delta)^2}{(x - i\Delta)^2} \qquad (C.1)
$$

$$
= \frac{x^2 - \Delta^2 - 2ix\Delta}{(x^2 + \Delta^2)^2} = \frac{1}{x^2 + \Delta^2} - \frac{2\Delta^2}{(x^2 + \Delta^2)^2} - \frac{2ix\Delta}{(x^2 + \Delta^2)^2} \qquad (C.2)
$$

$$
= \frac{A}{2\Delta} - \frac{1}{2}A^2 - \frac{ix}{2\Delta}A^2 \qquad (C.3)
$$

with the spectral function

$$
A = \frac{-2\Im\Sigma}{[E - H - \Re\Sigma]^2 + (\Im\Sigma)^2} = \frac{2\Delta}{x^2 + \Delta^2} \cdot \qquad (C.4)
$$

In the weak-disorder limit the impurity concentration $n_i \to 0$ and therefore $\Delta \to 0$. From Eq. (C.4) one can easily see that for $\Delta \to 0$ the spectral function becomes a δ-function. The application of this limiting procedure to

121

Eq. (C.3) leads to the following expression

$$\langle G^+\rangle\langle G^+\rangle = \frac{A}{2\Delta} - \frac{1}{2}A^2 - \frac{ix}{2\Delta}A^2 \to \frac{A}{2\Delta} - \frac{A}{2\Delta} - \frac{ix}{2\Delta^2}A \to -\frac{i}{2\Delta^2}x\delta(x) = 0$$
(C.5)

where the identity $A^2 \to A/\Delta$ for $\Delta \to 0$ [88] has been used.

If one investigates the behavior of the term which is proportional to $\langle G^+\rangle\langle G^-\rangle$ one obtains

$$\langle G^+\rangle\langle G^-\rangle = \frac{1}{(E - H - \Re\Sigma - i\Im\Sigma)} \times \frac{1}{(E - H - \Re\Sigma + i\Im\Sigma)}$$
(C.6)

$$= \frac{1}{[E - H - \Re\Sigma]^2 + [\Im\Sigma]^2} = \frac{A}{-2\Im\Sigma} = \frac{A}{2\Delta}.$$
(C.7)

The last equation clearly shows that the term $\langle G^+\rangle\langle G^-\rangle$ diverges if $\Delta \to 0$ implying that if the disorder approaches zero the conductivity becomes infinite.

The considerations in this appendix justify the approximation of the conductivity by Eq. (4.62) which is used for the diagrammatic representation of the conductivity in the weak-disorder limit.

Appendix D

Boltzmann Equation

The central quantity of the Boltzmann equation is the distribution function $f_\mathbf{k}(\mathbf{r})$. This distribution function gives the concentration of charge carriers in a quantum state $\mathbf{k}$ at position $\mathbf{r}$. This distribution function can change during time trough three main processes: diffusion, the influence of external fields and scattering. The Boltzmann equation states that the total rate of change has to vanish [2, 188]:

$$-\frac{\partial f_\mathbf{k}}{\partial t}\bigg|_{\text{scatt.}} + \frac{\partial f_\mathbf{k}}{\partial t}\bigg|_{\text{field}} + \frac{\partial f_\mathbf{k}}{\partial t}\bigg|_{\text{diff.}} = 0 . \tag{D.1}$$

Neglecting the diffusion term the Boltzmann equation consists of the external field term (here only an electric field $\mathbf{E}$ is considered) [189]:

$$\frac{\partial f_\mathbf{k}}{\partial t}\bigg|_{\text{field}} = \frac{\mathrm{d}\mathbf{k}}{\mathrm{d}t}\frac{\partial f_\mathbf{k}}{\partial E_\mathbf{k}}\frac{\partial E_\mathbf{k}}{\partial \mathbf{k}} = -|e|\frac{\partial f_\mathbf{k}}{\partial E_\mathbf{k}}\mathbf{v_k}\cdot\mathbf{E} , \tag{D.2}$$

with $\dot{\mathbf{k}} = -|e|\mathbf{E}$ and $E_\mathbf{k}$ are eigenvalues of the Bloch states of the ideal crystal and the scattering term. The change of the distribution function $f_\mathbf{k}$ due to scattering processes is given by [188]:

$$\frac{\partial f_\mathbf{k}}{\partial t}\bigg|_{\text{scatt.}} = \sum_{\mathbf{k}'}[f_{\mathbf{k}'}(1-f_\mathbf{k})P_{\mathbf{k}'\mathbf{k}} - (1-f_{\mathbf{k}'})f_\mathbf{k}P_{\mathbf{k}\mathbf{k}'}] , \tag{D.3}$$

where $P_{\mathbf{k}'\mathbf{k}}$ corresponds to a transition probability for scattering from state $\mathbf{k}'$ to state $\mathbf{k}$. Therefore, the first term from Eq. (D.3) describes the scattering from the occupied state $\mathbf{k}'$ to the unoccupied state $\mathbf{k}$. This term is the so-called "scattering-in" term which corresponds in the diluted limit to the vertex corrections of the Kubo-Středa equation [89]. The second term is the "scattering-out" term which describes the scattering from an occupied state

$\mathbf{k}$ into an unoccupied state $\mathbf{k}'$.

Using the Boltzmann equation in the spirit of linear response theory (small $\mathbf{E}$, $g_\mathbf{k} \propto \mathbf{E}$) one can assume that the deviation $g_\mathbf{k}$ of the equilibrium distribution function $f_\mathbf{k}^0$ (Fermi-Dirac distribution) is small:

$$f_\mathbf{k} = f_\mathbf{k}^0 + g_\mathbf{k} \ , \tag{D.4}$$

with $g_\mathbf{k} \ll f_\mathbf{k}^0$. This leads to:

$$\left. \frac{\partial f_\mathbf{k}}{\partial t} \right|_{\text{scatt.}} = \sum_{\mathbf{k}'} P_{\mathbf{k}\mathbf{k}'} \left(g_{\mathbf{k}'} - g_\mathbf{k} \right) , \tag{D.5}$$

where $P_{\mathbf{k}'\mathbf{k}} = P_{\mathbf{k}\mathbf{k}'}$ has been used. Making the ansatz [188]:

$$g_\mathbf{k} = -|e| \, \delta(E_\mathbf{k} - E_F) \, \boldsymbol{\Lambda}_\mathbf{k} \cdot \mathbf{E} \ , \tag{D.6}$$

where $\boldsymbol{\Lambda}_\mathbf{k}$ is the vector mean free path. Combining Eqs. (D.1), (D.2), (D.5) and (D.6) one obtains:

$$-|e| \frac{\partial f_\mathbf{k}}{\partial E_\mathbf{k}} \, \mathbf{v}_\mathbf{k} \cdot \mathbf{E} = - \, |e| \sum_{\mathbf{k}'} P_{\mathbf{k}\mathbf{k}'} \, [\delta(E_{\mathbf{k}'} - E_F) \, \boldsymbol{\Lambda}_{\mathbf{k}'} \cdot \mathbf{E} - \delta(E_\mathbf{k} - E_F) \, \boldsymbol{\Lambda}_\mathbf{k} \cdot \mathbf{E}] \tag{D.7}$$

$$-\delta(E_\mathbf{k} - E_F) \, \mathbf{v}_\mathbf{k} \cdot \mathbf{E} = \sum_{\mathbf{k}'} P_{\mathbf{k}\mathbf{k}'} \, [\delta(E_{\mathbf{k}'} - E_F) \, \boldsymbol{\Lambda}_{\mathbf{k}'} \cdot \mathbf{E}] - \tau_\mathbf{k}^{-1} \delta(E_\mathbf{k} - E_F) \, \boldsymbol{\Lambda}_\mathbf{k} \cdot \mathbf{E} \tag{D.8}$$

$$\boldsymbol{\Lambda}_\mathbf{k} \cdot \mathbf{E} \, \delta(E_\mathbf{k} - E_F) = \tau_\mathbf{k} \left\{ \mathbf{v}_\mathbf{k} \cdot \mathbf{E} \, \delta(E_\mathbf{k} - E_F) + \sum_{\mathbf{k}'} P_{\mathbf{k}\mathbf{k}'} \, [\boldsymbol{\Lambda}_{\mathbf{k}'} \cdot \mathbf{E} \, \delta(E_{\mathbf{k}'} - E_F)] \right\} \tag{D.9}$$

where higher order terms in $\mathbf{E}$ have been neglected and the Boltzmann relaxation time $\tau_\mathbf{k}^{-1} = \sum_{\mathbf{k}'} P_{\mathbf{k}\mathbf{k}'}$ has been used. From Eq. (D.9) one obtains for the vector mean free path:

$$\boldsymbol{\Lambda}_\mathbf{k} = \tau_\mathbf{k} \left(\mathbf{v}_\mathbf{k} + \sum_{\mathbf{k}'} P_{\mathbf{k}\mathbf{k}'} \boldsymbol{\Lambda}_{\mathbf{k}'} \right) \ . \tag{D.10}$$

The second term from this equation corresponds to the scattering-in term which causes the vertex corrections. In order to solve this equation an iterative scheme can be employed [190]. Finally, the conductivity can be calculated via a Fermi surface integral [189]:

$$\sigma_{xy} = \frac{e^2}{(2\pi)^3} \sum_n \iint_{\text{FS}} dS \, \frac{1}{v_\mathbf{k}^n} \, v_\mathbf{k}^{n,x} \, \Lambda_\mathbf{k}^{n,y} \ . \tag{D.11}$$

(here an additional band index n has been used). In order to calculate the extrinsic spin Hall effect which based on skew scattering one can use a similar equation [191]:

$$\sigma_{xy}^z = \frac{e^2}{(2\pi)^3} \sum_n \iint_{\mathrm{FS}} dS \, \frac{s_n^z(\mathbf{k})}{v_{\mathbf{k}}^n} \, v_{\mathbf{k}}^{n,x} \, \Lambda_{\mathbf{k}}^{n,y} \,, \tag{D.12}$$

with matrix elements $s_z^n(\mathbf{k}) = \langle \phi_{\mathbf{k}}^n | \beta \, \Sigma_z | \phi_{\mathbf{k}}^n \rangle$ (the Boltzmann formalism results shown in Fig. 7.16 are multiplied by -1 in order to be consistent with the conversion from spin conductivity into the unit of charge conductivity used within the present work). It turns out that only the scattering-in term gives a contribution to skew scattering. This observation is in line with the analysis of the contributions to skew scattering within the Kubo-Středa formalism. In Sec. 7.1 it was mentioned that only a certain class of vertex diagrams contribute to skew scattering.

Appendix E

Reformulation of the Spin-Current Density Operator Matrix Element

The matrix element of the current density operator $\hat{\mathbf{j}}$ has the following form:

$$-\langle \Lambda, E | ec\boldsymbol{\alpha} | \Lambda', E \rangle \,, \tag{E.1}$$

with $\Lambda = (\kappa, \mu)$ and E gives the energy dependence $(e = |e|)$. An exact reformulation of this matrix element for non-magnetic systems is [192]:

$$-\langle \Lambda, E | ec\boldsymbol{\alpha} | \Lambda', E \rangle = -\frac{e}{m + E/c^2} \langle \Lambda, E | \hat{\mathbf{p}} + \frac{V}{c}\boldsymbol{\alpha} | \Lambda', E \rangle \,, \tag{E.2}$$

with the canonical momentum operator $\hat{\mathbf{p}}$ and the potential V.

Vernes et al. [97] showed that a proper relativistic spin-current density operator can be defined as $\hat{J}_i^j = ec\alpha_i T_j$ where $\mathbf{T}$ is the spatial component of the Bargmann-Wigner four-vector polarization operator (see Sec. 5.2). Projecting $\mathbf{T}$ to the z-axis one arrives for non-magnetic systems at the following expression:

$$T_z = \beta\Sigma_z - \frac{1}{mc}\gamma_5\hat{p}_z \,. \tag{E.3}$$

This leads for a spin projection on the z-axis in combination with a current along the x-axis to the spin-current density operator:

$$\hat{J}_x^z = ec\left(\beta\Sigma_z\alpha_x - \frac{1}{mc}\gamma_5\hat{p}_z\alpha_x\right) \,. \tag{E.4}$$

In the following it will be shown that the matrix element of the first term can be reformulated in a similar way as shown by Eq. (E.2). In order to perform

127

such a reformulation one can use the relation:

$$\langle \Lambda, E | [H_{\mathrm{D}}, \beta \Sigma_z \alpha_x]_+ | \Lambda', E \rangle = 2(mc^2 + E)\langle \Lambda, E | \beta \Sigma_z \alpha_x | \Lambda', E \rangle \ , \qquad (E.5)$$

where H_{D} corresponds to the Dirac Hamiltonian. Calculating the anti-commutator $[H_{\mathrm{D}}, \beta \Sigma_z \alpha_x]_+$ for non-magnetic systems i.e.:

$$H_{\mathrm{D}} = c\boldsymbol{\alpha} \cdot \hat{\mathbf{p}} + \beta mc^2 + V \ , \qquad (E.6)$$

leads to the following expression:

$$
\begin{aligned}
[H_{\mathrm{D}}, \beta \Sigma_z \boldsymbol{\alpha}]_+ =& H_{\mathrm{D}} \beta \Sigma_z \boldsymbol{\alpha} + \beta \Sigma_z \boldsymbol{\alpha} H_{\mathrm{D}} && (E.7) \\
=& c\boldsymbol{\alpha} \cdot \hat{\mathbf{p}} \beta \Sigma_z \boldsymbol{\alpha} + \beta mc^2 \beta \Sigma_z \boldsymbol{\alpha} + V \beta \Sigma_z \boldsymbol{\alpha} + \\
&+ \beta \Sigma_z \boldsymbol{\alpha} c \boldsymbol{\alpha} \cdot \hat{\mathbf{p}} + \beta \Sigma_z \boldsymbol{\alpha} \beta mc^2 + \beta \Sigma_z \boldsymbol{\alpha} V && (E.8) \\
=& c\beta \Sigma_z [(\alpha_x \hat{p}_x + \alpha_y \hat{p}_y - \alpha_z \hat{p}_z)\boldsymbol{\alpha} + \boldsymbol{\alpha}\boldsymbol{\alpha} \cdot \hat{\mathbf{p}}] \\
&+ mc^2 \underbrace{(\Sigma_z \boldsymbol{\alpha} + \beta \Sigma_z \boldsymbol{\alpha} \beta)}_{=0} + 2V\beta \Sigma_z \boldsymbol{\alpha} && (E.9) \\
=& c\beta \Sigma_z(\boldsymbol{\alpha} \cdot \hat{\mathbf{p}} \boldsymbol{\alpha} + \boldsymbol{\alpha}\boldsymbol{\alpha} \cdot \hat{\mathbf{p}}) - 2c\beta \Sigma_z \alpha_z \hat{p}_z \boldsymbol{\alpha} + 2V\beta \Sigma_z \boldsymbol{\alpha} && (E.10) \\
=& 2c\beta \Sigma_z(\hat{\mathbf{p}} + \frac{V}{c}\boldsymbol{\alpha}) + 2c\beta \gamma_5 \hat{p}_z \boldsymbol{\alpha} \ , && (E.11)
\end{aligned}
$$

where it has been used that $\alpha_i \beta = -\beta \alpha_i$, $[\alpha_i, \Sigma_z]_+ = -2\gamma_5 \delta_{iz}$ with $i = x, y, z$ and $[\alpha_i, \alpha_j] = 2\alpha_i \alpha_j (1 - \delta_{ij})$. Inserting Eq. (E.11) into Eq. (E.5) leads to:

$$\langle \Lambda, E | \beta \Sigma_z \alpha_x | \Lambda', E \rangle = \frac{1}{mc + E/c} \langle \Lambda, E | [\beta \Sigma_z(\hat{p}_x + \frac{V}{c}\alpha_x) + \beta \gamma_5 \hat{p}_z \alpha_x] | \Lambda', E \rangle \ . \qquad (E.12)$$

Therefore, the matrix element of the spin-current density operator shown by Eq. (E.4) transforms to:

$$
\begin{aligned}
\langle \Lambda, E | \hat{J}_x^z | \Lambda', E \rangle =& \frac{e}{m + E/c^2} \langle \Lambda, E | \beta \Sigma_z(\hat{p}_x + \frac{V}{c}\alpha_x) | \Lambda', E \rangle \\
&+ e\langle \Lambda, E | \frac{1}{m + E/c^2} \beta \gamma_5 \hat{p}_z \alpha_x - \frac{1}{m} \gamma_5 \hat{p}_z \alpha_x | \Lambda', E \rangle && (E.13) \\
=& \frac{e}{m + E/c^2} \langle \Lambda, E | \beta \Sigma_z(\hat{p}_x + \frac{V}{c}\alpha_x) | \Lambda', E \rangle \\
&+ e\langle \Lambda, E | - \frac{1}{m + E/c^2} \beta \Sigma_x \hat{p}_z + \frac{1}{m} \Sigma_x \hat{p}_z | \Lambda', E \rangle && (E.14) \\
=& \frac{e}{m + E/c^2} \langle \Lambda, E | \beta \Sigma_z(\hat{p}_x + \frac{V}{c}\alpha_x) | \Lambda', E \rangle \\
&+ \frac{e}{m + E/c^2} \langle \Lambda, E | [(1 + \frac{E}{mc^2})\mathbb{1}_4 - \beta] \Sigma_x \hat{p}_z | \Lambda', E \rangle \ . && (E.15)
\end{aligned}
$$

Due to the fact that $E/(mc^2)$ is small and β affects only the contribution due to the small components of the bi-spinors the second term from Eq. (E.15) is negligible. This leads to the following form of the spin-current density operator matrix element:

$$\langle \Lambda, E | \hat{J}_x^z | \Lambda', E \rangle \approx \frac{e}{m + E/c^2} \langle \Lambda, E | \beta \Sigma_z (\hat{p}_x + \frac{V}{c}\alpha_x) | \Lambda', E \rangle , \qquad (E.16)$$

which can be used for the calculation of the SHC. The advantage of the matrix element shown in Eq. (E.16) compared to the matrix element of the operator given by Eq. (E.4) is that the dominant contribution to the SHC comes from the first term in Eq. (E.16) which allows to neglect the term which includes the α_μ matrix. Due to the fact that the α_μ matrix couples the small and large component of the bi-spinors whereas $\hat{p}$ couples the large and small components independently (what allows to consider the large component only) the matrix element is drastically simplified. However, it is important to note that effects like the microscopical side-jump scattering mechanism which is related to the anomalous velocity are no longer included in calculations which use the matrix element given by Eq. (E.16) neglecting the potential term.

Appendix E. Reformulation of the Spin-Current Density Operator Matrix Element

Appendix F

Computational Details

All calculations were done in the framework of spin density functional theory using the local spin density approximation (LSDA) using the parameterization of Vosko, Wilk and Nusair [36] for the exchange correlation functional. The potential construction has been done within the atomic sphere approximation (ASA). All transport results include vertex corrections if not stated otherwise. For the angular momentum expansion a cutoff of $l_{\mathrm{max}} = 3$ has been used apart from the results for $Ga_{1-x}Mn_xAs$ for which $l_{\mathrm{max}} = 2$ has been used. A very crucial point within the presented transport calculations is the number of k-points used for the Brillouin zone integration. Therefore, well converged k-meshes in the range of ≈ 50000 (residual resistivity) up to 10^9 (very low impurity concentrations AHE) k-points within the first Brillouin zone have been used.

Appendix F. Computational Details

Appendix G

Acronyms

The present work uses a variety of acronyms where the most important are listed in the following:

- AHC   anomalous Hall conductivity
- AHE   anomalous Hall effect
- AMR   anisotropic magnetoresistance
- BSF   Bloch spectral function
- CPA   coherent potential approximation
- DFT   density functional theory
- DOS   density of states
- GF   Green's function
- KKR   Korringa-Kohn-Rostoker
- NLCPA   non-local coherent potential approximation
- OHE   ordinary Hall effect
- OrbHC   orbital Hall conductivity
- OrbHE   orbital Hall effect
- SHC   spin Hall conductivity
- SHE   spin Hall effect
- VC   vertex corrections

Appendix G. Acronyms

Bibliography

[1] L. Brillouin, *Wave Propagation in Periodic Structures* (Dover, New York, 1953).

[2] N. Ashcroft and N. Mermin, *Solid State Physics* (Saunders College Publishers, New York, 1976).

[3] P. L. Rossiter, *The Electrical Resistivity of Metals and Alloys* (Cambridge University Press, Cambridge, 1987).

[4] G. A. Prinz, "Magnetoelectronics," Science **282**, 1660 (1998).

[5] S. A. Wolf, D. D. Awschalom, R. A. Buhrman, J. M. Daughton, S. von Molnár, M. L. Roukes, A. Y. Chtchelkanova, and D. M. Treger, "Spintronics: A spin-based electronics vision for the future," Science **294**, 1488 (2001).

[6] I. Žutić, J. Fabian, and S. Das Sarma, "Spintronics: Fundamentals and applications," Rev. Mod. Phys. **76**, 323 (2004).

[7] D. D. Awschalom and M. Flatté, "Challenges for semiconductor spintronics," Nature Phys. **3**, 153 (2007).

[8] D. Awschalom and N. Samarth, "Spintronics without magnetism," Physics **2**, 50 (2009).

[9] M. N. Baibich, J. M. Broto, A. Fert, F. N. V. Dau, F. Petroff, P. Etienne, G. Creuzet, A. Friederich, and J. Chazelas, "Giant magnetoresistance of (001)Fe/(001)Cr magnetic superlattices," Phys. Rev. Letters **61**, 2472 (1988).

[10] G. Binasch, P. Grünberg, F. Saurenbach, and W. Zinn, "Enhanced magnetoresistance in layered magnetic structures with antiferromagnetic interlayer exchange," Phys. Rev. B **39**, 4828 (1989).

[11] P. A. Grünberg, "Nobel lecture: From spin waves to giant magnetoresistance and beyond," Rev. Mod. Phys. **80**, 1531 (2008).

[12] J. L. Cheng and M. W. Wu, "Kinetic investigation of the extrinsic spin Hall effect induced by skew scattering," J. Phys.: Condensed Matter **20**, 085209 (2008).

[13] T. Kimura, Y. Otani, T. Sato, S. Takahashi, and S. Maekawa, "Room-temperature reversible spin Hall effect," Phys. Rev. Letters **98**, 156601 (2007).

[14] M. I. Dyakonov and V. I. Perel, "Possibility of orientating electron spins with current," JETP Lett. **13**, 467 (1971).

[15] M. I. Dyakonov and V. I. Perel, "Current-induced spin orientation of electrons in semiconductors," Phys. Lett. A **35**, 459 (1971).

[16] J. E. Hirsch, "Spin Hall effect," Phys. Rev. Letters **83**, 1834 (1999).

[17] J. Inoue and H. Ohno, "Taking the Hall effect for a spin," Science **309**, 2004 (2005).

[18] L. H. Thomas, "The calculation of atomic fields," Proc. Cambridge Phil. Soc. **23**, 542 (1927).

[19] E. Fermi, "Un metodo statistice per la determinazione di alcune proprieta dell'atomo," Rend. Accad. Naz. Linzei **6**, 602 (1927).

[20] P. Hohenberg and W. Kohn, "Inhomogenous electron gas," Phys. Rev. **136**, B 864 (1964).

[21] R. G. Parr and W. Yang, *Density-Functional Theory of Atoms and Molecules* (Oxford University Press, New York, 1989).

[22] M. Levy, "Universal variational functionals of electron densities, first-order density matrices, and natural spin-orbitals and solution of the v-representability problem," Proc. Natl. Acad. Sci. USA **76** (1979).

[23] C. Cohen-Tannoudji, B. Diu, and F. Laloë, *Quantum Mechanics* (Wiley & Sons, 1977).

[24] J. P. Perdew and S. Kurth, in *A Primer in Density Functional Theory*, edited by C. Fiolhais, F. Nogueira, and M. Marques (Springer, Berlin, 2003).

[25] P. A. M. Dirac, "Note on exchange phenomena in the Thomas atom," Proc. Cambridge Phil. Soc. **26**, 376 (1930).

[26] W. Kohn and L. J. Sham, "Self-consistent equations including exchange and correlation effects," Phys. Rev. **140**, A 1133 (1965).

[27] A. K. Rajagopal and J. Callaway, "Inhomogeneous electron gas," Phys. Rev. B **7**, 1912 (1973).

[28] M. V. Ramana and A. K. Rajagopal, "Relativistic spin-polarised electron gas," J. Phys. C: Solid State Phys. **12**, L845 (1979).

[29] H. Eschrig, G. Seifert, and P. Ziesche, "Current density functional theory of quantum electrodynamics," Solid State Commun. **56**, 777 (1985).

[30] R. M. Dreizler and E. K. U. Gross, *Density Functional Theory* (Springer-Verlag, Heidelberg, 1990).

[31] M. E. Rose, *Relativistic Electron Theory* (Wiley, New York, 1961).

[32] H. Eschrig, *The Fundamentals of Density Functional Theory* (B G Teubner Verlagsgesellschaft, Stuttgart, Leipzig, 1996).

[33] E. Engel, T. Auth, and R. Dreizler, "Relativistic spin-density-functional theory: Robust solution of single-particle equations for open-subshell atoms," Phys. Rev. B **64**, 235126/1 (2001).

[34] A. H. MacDonald and S. H. Vosko, "A relativistic density functional formalism," J. Phys. C: Solid State Phys. **12**, 2977 (1979).

[35] W. Koch and M. C. Holthausen, *A Chemist's Guide to Density Functional Theory* (Wiley-VCH, 2001).

[36] S. H. Vosko, L. Wilk, and M. Nusair, "Accurate spin-dependent electron liquid correlation energies for local spin density calculations: a critical analysis," Can. J. Phys. **58**, 1200 (1980).

[37] A. D. Becke, "Density-functional exchange-energy approximation with correct asymptotic behavior," Phys. Rev. A **38**, 3098 (1988).

[38] J. P. Perdew and Y. Wang, "Accurate and simple analytic representation of the electron-gas correlation energy," Phys. Rev. B **45**, 13244 (1992).

[39] J. Perdew, K. Burke, and M. Ernzerhof, "Generalized gradient approximation made simple," Physical Review Letters **77**, 3865 (1996).

[40] L. Rayleigh, "On the influence of obstacles arranged in rectangular order upon the properties of a medium," Phil. Mag. **34**, 481 (1892).

[41] A. Gonis and W. H. Butler, *Multiple Scattering in Solids* (Springer, New York, 2000).

[42] J. Korringa, "On the calculation of the energy of a bloch wave in a metal," Physica **XIII**, 392 (1947).

[43] W. Kohn and N. Rostoker, "Solution of the Schrödinger equation in periodic lattices with an application to metallic lithium," Phys. Rev. **94**, 1111 (1954).

[44] T. H. Dupree, "Electron scattering in a crystal lattice," Ann. Phys. (New York) **15**, 63 (1961).

[45] J. L. Beeby, "The density of electrons in a perfect or imperfect lattice," Proc. Roy. Soc. (London) A **302**, 113 (1967).

[46] N. A. W. Holzwarth, "Theory of impurity scattering in dilute metal alloys based on the muffin-tin model," Phys. Rev. B **11**, 3718 (1975).

[47] R. Feder, F. Rosicky, and B. Ackermann, "Relativistic multiple scattering theory of electrons by ferromagnets," Z. Physik B **52**, 31 (1983).

[48] P. Strange, J. B. Staunton, and B. L. Györffy, "Relativistic spin-polarized scattering theory – solution of the single-site problem," J. Phys. C: Solid State Phys. **17**, 3355 (1984).

[49] K. Elk and W. Gasser, *Die Methode der Greenschen Funktionen in der Festkörperphysik* (Akademie-Verlag, Berlin, 1979).

[50] E. N. Economou, *Green's Functions in Quantum Physics* (Springer-Verlag, New York, 2006).

[51] I. N. Bronstein, K. A. Semendjajew, G. Musiol, and H. Mühlig, *Taschenbuch der Mathematik* (Verlag Harri Deutsch, Frankfurt a. M., 2000).

[52] P. Strange, *Relativistic Quantum Mechanics* (Cambridge University Press, Cambridge, 1998).

[53] J. Zabloudil, R. Hammerling, L. Szunyogh, and P. Weinberger, *Electron Scattering in Solid Matter* (Springer, Berlin, 2005).

[54] E. Tamura, "Relativistic single-site Green function for general potentials," Phys. Rev. B **45**, 3271 (1992).

[55] R. G. Newton, *Scattering Theory of Waves and Particles* (McGraw-Hill, New York, 1967).

[56] H. Ebert, Habilitation thesis, University of München (1990).

[57] A. C. Jenkins and P. Strange, "Relativistic spin-polarized single-site scattering theory," J. Phys.: Condensed Matter **6**, 3499 (1994).

[58] B. L. Györffy and M. J. Stott, *Band Structure Spectroscopy of Metals and Alloys* (Academic Press, New York, 1973), p. 385.

[59] J. S. Faulkner and G. M. Stocks, "Calculating properties with the coherent-potential approximation," Phys. Rev. B **21**, 3222 (1980).

[60] A. Gonis, *Green Functions for Ordered and Disordered Systems* (North-Holland, Amsterdam, 1992).

[61] R. J. Elliott, J. A. Krumhansl, and P. L. Leath, "The theory and properties of randomly disordered crystals and related physical systems," Rev. Mod. Phys. **46**, 465 (1974).

[62] P. Soven, "Coherent-potential model of substitutional disordered alloys," Phys. Rev. **156**, 809 (1967).

[63] J. Rammer, *Quantum Transport Theory* (Westview Press, 2004).

[64] W. Nolting, *Grundkurs Theoretische Physik 7 (Viel-Teilchen-Theorie)* (Vieweg, Braunschweig, 1990).

[65] D. Binosi and L. Theußl, "JaxoDraw: A graphical user interface for drawing Feynman diagrams," Comp. Phys. Comm. **161**, 76 (2004).

[66] D. A. Rowlands, J. B. Staunton, and B. L. Györffy, "Korringa-Kohn-Rostoker nonlocal coherent-potential approximation," Phys. Rev. B **67**, 115109 (2003).

[67] D. A. Biava, S. Ghosh, D. D. Johnson, W. A. Shelton, and A. V. Smirnov, "Systematic, multisite short-range-order corrections to the electronic structure of disordered alloys from first principles: The KKR nonlocal CPA from the dynamical cluster approximation," Phys. Rev. B **72**, 113105 (2005).

[68] D. A. Rowlands, X.-G. Zhang, and A. Gonis, "Reformulation of the nonlocal coherent-potential approximation as a unique reciprocal-space theory of disorder," Phys. Rev. B **78**, 115119 (2008).

[69] D. A. Rowlands, "Short-range correlations in disordered systems: non-local coherent-potential approximation," Rep. Prog. Phys. **72**, 086501 (2009).

[70] M. H. Hettler, A. N. Tahvildar-Zadeh, M. Jarrell, T. Pruschke, and H. R. Krishnamurthy, "Nonlocal dynamical correlations of strongly interacting electron systems," Phys. Rev. B **58**, R7475 (1998).

[71] M. Jarrell and H. R. Krishnamurthy, "Systematic and causal corrections to the coherent potential approximation," Phys. Rev. B **63**, 125102 (2001).

[72] D. Ködderitzsch, H. Ebert, D. A. Rowlands, and A. Ernst, "Relativistic formulation of the Korringa-Kohn-Rostoker non-local coherent-potential approximation," New Journal of Physics **9**, 81 (2007).

[73] D. A. Rowlands, J. B. Staunton, B. L. Györffy, E. Bruno, and B. Ginatempo, "Effects of short-range order on the electronic structure of disordered metallic systems," Phys. Rev. B **72**, 045101 (2005).

[74] G. M. Stocks and W. H. Butler, "Mass and lifetime enhancement due to disorder on Ag_cPd_{1-c} alloys," Phys. Rev. Letters **48**, 55 (1982).

[75] J. S. Faulkner, "The modern theory of alloys," Prog. Mater. Sci. **27**, 3 (1982).

[76] P. R. Tulip, J. B. Staunton, D. A. Rowlands, B. L. Györffy, E. Bruno, and B. Ginatempo, "Nonsite diagonal properties from the Korringa-Kohn-Rostoker nonlocal coherent-potential approximation," Phys. Rev. B **73**, 205109 (2006).

[77] M. S. Green, "Markoff random processes and the statistical mechanics of time-dependent phenomena," J. Chem. Phys. **20**, 1281 (1952).

[78] R. Kubo, "Statistical-mechanical theory of irreversible processes. I. General theory and simple application to magnetic and conduction problems," J. Phys. Soc. Japan **12**, 570 (1957).

[79] P. Středa, "Theory of quantised Hall conductivity in two dimensions," J. Phys. C: Solid State Phys. **15**, L717 (1982).

[80] A. Bastin, C. Lewiner, O. Betbeder-Matibet, and P.Nozieres, "Quantum oscillations of the Hall effect of a fermion gas with random impurity scattering," J. Phys. Chem. Solids **32**, 1811 (1971).

[81] A. Crépieux and P. Bruno, "Theory of the anomalous Hall effect from the Kubo formula and the Dirac equation," Phys. Rev. B **64**, 014416/1 (2001).

[82] Y. Yao, L. Kleinman, A. H. MacDonald, J. Sinova, T. Jungwirth, D. Wang, E. Wang, and Q. Niu, "First principles calculation of anomalous Hall conductivity in ferromagnetic bcc Fe," Phys. Rev. Letters **92**, 037204 (2004).

[83] J. Sinova, D. Culcer, Q. Niu, N. A. Sinitsyn, T. Jungwirth, and A. H. MacDonald, "Universal intrinsic spin Hall effect," Phys. Rev. Letters **92**, 126603 (2004).

[84] G. Y. Guo, Y. Yao, and Q. Niu, "*Ab initio* calculation of the intrinsic spin Hall effect in semiconductors," Phys. Rev. Letters **94**, 226601 (2005).

[85] I. A. Campbell and A. Fert, *Ferromagnetic Materials* (North-Holland, Amsterdam, 1982), vol. 3, p. 751.

[86] D. A. Greenwood, "The Boltzmann equation in the theory of the electrical conduction of metals," Proc. Phys. Soc. **71**, 585 (1958).

[87] N. A. Sinitsyn, A. H. MacDonald, T. Jungwirth, V. K. Dugaev, and J. Sinova, "Anomalous Hall effect in a two-dimensional Dirac band: The link between the Kubo-Streda formula and the semiclassical Boltzmann equation approach," Phys. Rev. B **75**, 045315 (2007).

[88] G. D. Mahan, *Many-Particle Physics* (Springer, Netherlands, 2000).

[89] W. H. Butler, "Theory of electronic transport in random alloys: Korringa-Kohn-Rostoker coherent-potential approximation," Phys. Rev. B **31**, 3260 (1985).

[90] V. K. Dugaev, A. Crépieux, and P. Bruno, "Localization corrections to the anomalous Hall effect in a ferromagnet," Phys. Rev. B **64**, 104411 (2001).

[91] J. Banhart and H. Ebert, "First-principles calculation of spontaneous magnetoresistance anisotropy and anomalous Hall effect in disordered ferromagnetic alloys," Europhys. Lett. **32**, 517 (1995).

[92] P. R. Tulip, J. B. Staunton, S. Lowitzer, D. Ködderitzsch, and H. Ebert, "Theory of electronic transport in random alloys with short-range order: Korringa-Kohn-Rostoker non-local coherent potential approximation approach," Phys. Rev. B **77**, 165116 (2008).

[93] P. Grünberg, "It's the coupling that creates resistance: Spin electronics in layered magnetic structures," Ann. Physik **17**, 7 (2008).

[94] H. Ebert, A. Vernes, and J. Banhart, "Anisotropic electrical resistivity of ferromagnetic Co-Pd and Co-Pt alloys," Phys. Rev. B **54**, 8479 (1996).

[95] J. Banhart, H. Ebert, and A. Vernes, "Applicability of the two-current model for systems with strongly spin-dependent disorder," Phys. Rev. B **56**, 10165 (1997).

[96] T. Tanaka, H. Kontani, M. Naito, T. Naito, D. S. Hirashima, K. Yamada, and J. Inoue, "Intrinsic spin Hall effect and orbital Hall effect in $4d$ and $5d$ transition metals," Phys. Rev. B **77**, 165117 (2008).

[97] A. Vernes, B. L. Györffy, and P. Weinberger, "Spin currents, spin-transfer torque, and spin-Hall effects in relativistic quantum mechanics," Phys. Rev. B **76**, 012408 (2007).

[98] V. Bargmann and E. P. Wigner, "Group theoretical discussion of relativistic wave equations," Proc. Natl. Acad. Sci. U.S.A. **34**, 211 (1948).

[99] V. Popescu, H. Ebert, N. Papanikolaou, R. Zeller, and P. H. Dederichs, "Spin-dependent transport in ferromagnet/semiconductor/ferromagnet junctions: A fully relativistic approach," J. Phys.: Condensed Matter **16**, S5579 (2004).

[100] D. M. Fradkin and R. H. Good, "Electron polarization operators," Rev. Mod. Phys. **33**, 343 (1961).

[101] S. Lowitzer, D. Ködderitzsch, H. Ebert, and J. B. Staunton, "Electronic transport in ferromagnetic alloys and the Slater-Pauling curve," Phys. Rev. B **79**, 115109 (2009).

[102] I. Mertig, R. Zeller, and P. H. Dederichs, "Ab initio calculations of residual resistivities for dilute Ni alloys," Phys. Rev. B **47**, 16178 (1993).

[103] A. Fert, "Nobel lecture: Origin, development, and future of spintronics," Rev. Mod. Phys. **80**, 1517 (2008).

[104] J. W. F. Dorleijn and A. R. Miedema, "A quantitative investigation of the two current conduction in nickel alloys," J. Phys. F: Met. Phys. **5**, 487 (1975).

[105] T. Jungwirth, J. Sinova, J. Mašek, J. Kučera, and A. H. MacDonald, "Theory of ferromagnetic (III,Mn)V semiconductors," Rev. Mod. Phys. **78**, 809 (2006).

[106] V. Novák, K. Olejník., J. Wunderlich, M. Cukr, K. Výborný, A. W. Rushforth, K. W. Edmonds, R. P. Campion, B. L. Gallagher, J. Sinova, et al., "Curie point singularity in the temperature derivative of resistivity in (Ga,Mn)As," Phys. Rev. Letters **101**, 077201 (2008).

[107] K. Yu, W. Walukiewicz, T. Wojtowicz, I. Kuryliszyn, X. Liu, Y. Sasaki, and J. Furdyna, "Effect of the location of Mn sites in ferromagnetic $Ga_{1-x}Mn_xAs$ on its Curie temperature," Phys. Rev. B **65**, 201303 (2002).

[108] I. Turek, J. Kudrnovksý, V. Drchal, and P. Weinberger, "Residual resistivity of diluted III-V magnetic semiconductors," J. Phys.: Condensed Matter **16**, S5607 (2004).

[109] K. W. Edmonds, K. Y. Wang, R. P. Campion, A. C. Neumann, C. T. Foxon, B. L. Gallagher, and P. C. Main, "Hall effect and hole densities in $Ga_{1-x}Mn_xAs$," Appl. Physics Lett. **81**, 3010 (2002).

[110] H. K. Choi, Y. S. Kim, S. S. A. Seo, I. T. Jeong, W. O. Lee, Y. S. Oh, K. H. Kim, J. C. Woo, T. W. Noh, Z. G. Khim, et al., "Evidence of metallic clustering in annealed $Ga_{1-x}Mn_xAs$ from atypical scaling behavior of the anomalous Hall coefficient," Appl. Physics Lett. **89**, 102503 (2006).

[111] S. H. Chun, Y. S., Kim, H. K. Choi, I. T. Jeong, W. O. Lee, K. S. Suh, Y. S. Oh, K. H. Kim, Z. G. Khim, et al., "Interplay between carrier and impurity concentrations in annealed $Ga_{1-x}Mn_xAs$: Intrinsic anomalous Hall effect," Phys. Rev. Letters **98**, 026601 (2007).

[112] F. Máca and J. Mašek, "Electronic states in $Ga_{1-x}Mn_xAs$: Substitutional versus interstitial position of Mn," Phys. Rev. B **65**, 235209/1 (2002).

[113] M. Migschitz, W. Garlipp, and W. Pfeiler, "Short-range order kinetics in α-AgZn for various states of post-deformation defect annealing after cold-rolling," Acta Metallurgica **44**, 2831 (1996).

[114] H. Thomas, "Über Widerstandslegierungen," Z. Phys. **129**, 219 (1951).

[115] T. Massalski, *Binary Alloy Phase Diagrams* (ASM International, Ohio, 1990).

[116] C. B. Walker and D. T. Keating, "Neutron diffraction study of short-range order in β-CuZn," Phys. Rev. **130**, 1726 (1963).

[117] L. Nordheim, "Zur Elektronentheorie der Metalle. II," Ann. Physik **9**, 641 (1931).

[118] B. E. Warren, *X-Ray Diffraction* (Addison-Wesley, London, UK, 1969).

[119] J. M. Cowley, "An approximate theory of order in alloys," Phys. Rev. **77**, 669 (1950).

[120] W. Webb, "A study of beta-brass in single crystal form," Phys. Rev. **55**, 297 (1938).

[121] R. M. Bozorth, *Ferromagnetism* (D. van Nostrand Company, New York, 1951).

[122] R. M. Bozorth, "Atomic moments of ferromagnetic alloys," Phys. Rev. (1950).

[123] N. F. Mott, "Electrons in transition metals," Adv. Phys. **13**, 325 (1964).

[124] J. Kübler, *Theory of Itinerant Electron Magnetism* (Oxford University Press, Clarendon, 2000).

[125] J. Staunton, "The electronic structure of magnetic transition metallic materials," Rep. Prog. Phys. **57**, 1289 (1994).

[126] A. P. Malozemoff, A. R. Williiams, and V. L. Moruzzi, "Band-gap theory of strong ferromagnetism: Application to concentrated crystalline and amorphous Fe- and Co-metalloid alloys," Phys. Rev. B **29**, 1620 (1984).

[127] D. D. Johnson, F. J. Pinski, and J. B. Staunton, "The Slater-Pauling curve: First principles calculations of the moments of $Fe_{1-c}Ni_c$ and $V_{1-c}Fe_c$," J. Appl. Physics **61**, 3715 (1987).

[128] Y. Y. Tsiovkin, A. N. Voloshinskii, V. V. Gaponstev, V. V. Ustinov, A. G. Obykhov, A. L. Nikolaev, I. A. Nekrasov, and A. V. Lukoyanov, "Anomalous concentration dependence of residual electrical resistivity in Fe-Cr alloys," Phys. Rev. B **72**, 224204 (2005).

[129] I. Mirebeau, M. Hennion, and G. Parette, "First measurement of short-range-order inversion as a function on concentration in a transition alloy," Phys. Rev. Lett. (1984).

[130] J. Cieślak, S. Dubiel, and B. Sepiol, "Mössbauer-effect study of the phase separation in the Fe-Cr system," J. Phys.: Condensed Matter **12**, 6709 (2000).

[131] P. Olsson, I. A. Abrikosov, and J. Wallenius, "Electronic origin of the anomalous stability of Fe-rich bcc Fe-Cr alloys," Phys. Rev. B **73**, 104416 (2006).

[132] A. L. Nikolaev, "Stage I of recovery in 5 MeV electron-irradiation iron and iron-chromium alloys: The effect of small cascades, migration of di-interstitials and mixed dumbbells," J. Phys.: Condensed Matter **11**, 8633 (1999).

[133] H. Ebert, A. Vernes, and J. Banhart, "Relativistic bandstructure of disordered magnetic alloys," Solid State Commun. **104**, 243 (1997).

[134] S. U. Jen and S. A. Chang, "Magnetic, thermal, and transport properties of $Fe_{100-x}V_x$ and $(Fe_{100-x}V_x)_{83}B_{17}$ alloys," Phys. Rev. B **47**, 5822 (1993).

[135] P. E. A. Turchi, L. Reinhard, and G. M. Stocks, "First-principles study of stability and local order in bcc-based Fe-Cr and Fe-V alloys," Phys. Rev. B **50**, 15542 (1994).

[136] R. E. Reed-Hill, *Physical Metallurgy Principles* (Van Nostrand, New York, 1972).

[137] P. Haasen, *Physical Metallurgy* (Cambridge University Press, Cambridge, 1996).

[138] R. Nordheim and N. J. Grant, "Resistivity anomalies in the nickel chromium system as evidence of ordering reactions," J. Inst. Metals **82**, 440 (1954).

[139] H. Baer, "Nahordnung und K-Zustand," Z. Metallkunde **56**, 79 (1965).

[140] F. Pařizek and A. Čižek, "The K-state in the Ni-Co alloy," Czech. J. Phys. B **19**, 276 (1969).

[141] T. S. Lei, K. Vasudevan, and E. E. Standbury, "Resistivity of short-range and long-range order changes in nickel-molybdenum (Ni_4Mo)," Mater. Res. Soc. Symp. Proc. **39**, 163 (1985).

[142] D. M. C. Nicholson and R. H. Brown, "Electrical resistivity of $Ni_{0.8}Mo_{0.2}$: Explanation of anomalous behavior in short-range ordered alloys," Phys. Rev. Letters **70**, 3311 (1993).

[143] J. H. Mooij, "Electrical conduction in concentrated disordered transition metal alloys," phys. stat. sol. (a) **17**, 521 (1973).

[144] A. M. Rana, A. F. Khan, A. Abbas, and M. I. Ansari, "Electrical resistivity behavior in Ni-25 at.% Cr alloy," Mater. Chem. Phys. **80**, 228 (2002).

[145] A. P. Druzhkov, V. P. Kolotushkin, V. L. Arbuzov, S. E. Danilov, and D. A. Perminov, "Structural and phase states and irradiation-induced defects in Ni-Cr alloys," Phys. Met. Metalloved. **101**, 369 (2005).

[146] D. A. Rowlands, J. B. Staunton, B. L. Györffy, E. Bruno, and B. Ginatempo, "Effects of short-range order on the electronic structure of disordered metallic systems," Phys. Rev. B **72**, 045101 (2005).

[147] J. E. Spruiell and E. E. Stansbury, "X-ray study of short-range order in nickel alloys containing 10.7 and 20.0 at. % molybdenum," J. Phys. Chem. Solids **26**, 811 (1965).

[148] B. Chakravarti, E. A. Starke, C. J. Sparks, and R. O. Williams, "Short range order and the development of long range order in Ni_4Mo," J. Phys. Chem. Solids **35**, 1317 (1974).

[149] A. Marucco and B. Nath, "Effects of ordering on the properties of Ni-Cr alloys," J. Mater. Sci. **1988**, 2107 (1988).

[150] Guénault, "Low-temperature thermoelectric power of palladium-silver alloys," Phil. Mag. **30**, 641 (1974).

[151] E. H. Hall, "On a new action of the magnet on electric currents," Amer. J. Math. **2**, 287 (1879).

[152] E. H. Hall, "On the rotational coefficient in nickel and cobalt," Phil. Mag. **12**, 157 (1881).

[153] N. A. Sinitsyn, "Semiclassical theories of the anomalous Hall effect," J. Phys.: Condensed Matter **20**, 023201 (2008).

[154] S. O. Valenzuela and M. Tinkham, "Direct electronic measurement of the spin Hall effect," Nature **442**, 176 (2006).

[155] E. Saitoh, M. Ueda, H. Miyajima, and G. Tatara, "Conversion of spin current into charge current at room temperature: Inverse spin-Hall effect," Appl. Physics Lett. **88**, 182509 (2006).

[156] L. Vila, T. Kimura, and Y. Otani, "Evolution of the spin Hall effect in Pt nanowires: Size and temperature effects," Phys. Rev. Letters **99**, 226604 (2007).

[157] T. Seki, Y. Hasegawa, S. Mitani, S. Takahashi, H. Imamura, S. Maekawa, J. Nitta, and K. Takanashi, "Giant spin Hall effect in perpendicularly spin-polarized FePt/Au devices," Nature Mater. **7**, 125 (2008).

[158] J. Smit, "The spontaneous Hall effect in ferromagnetics I," Physica **21**, 877 (1955).

[159] J. Smit, "The spontaneous Hall effect in ferromagnetics II," Physica **24**, 39 (1958).

[160] L. Berger, "Side-jump mechanism for the Hall effect in ferromagnets," Phys. Rev. B **2**, 4559 (1970).

[161] T. Jungwirth, Q. Niu, and A. H. MacDonald, "Anomalous Hall effect in ferromagnetic semiconductors," Phys. Rev. Letters **88**, 207208 (2002).

[162] N. Nagaosa, J. Sinova, S. Onoda, A. H. MacDonald, and N. P. Ong, "Anomalous Hall effect," Rev. Mod. Phys. **82**, 1539 (2010).

[163] Z. Fang, N. Nagaosa, K. S. Takahashi, A. Asamitsu, R. Mathieu, T. Ogasawara, H. Yamada, M. Kawasaki, Y. Tokura, and K. Terakura, "The anomalous Hall effect and magnetic monopoles in momentum space," Science **302**, 92 (2003).

[164] X. Wang, D. Vanderbilt, J. R. Yates, and I. Souza, "Fermi-surface calculation of the anomalous Hall conductivity," Phys. Rev. B **76**, 195109 (2007).

[165] Y. Yao and Z. Fang, "Sign change of intrinsic spin Hall effect in semiconductors and simple metals: First-principle calculations," Phys. Rev. Letters **95**, 156601 (2005).

[166] G. Y. Guo, S. Murkami, T.-W. Chen, and N. Nagaosa, "Intrinsic spin Hall effect in platinum: First-principle calculations," Phys. Rev. Letters **100**, 096401 (2008).

[167] G. Y. Guo, S. Maekawa, and N. Nagaosa, "Enhanced spin Hall effect by resonant skew scattering in the orbital-dependent Kondo effect," Phys. Rev. Letters **102**, 036401 (2009).

[168] G. Y. Guo, "*Ab initio* calculation of intrinsic spin Hall conductivity of Pd and Au," J. Appl. Physics **105**, 07C701 (2009).

[169] H. Kontani, T. Tanaka, and K. Yamada, "Intrinsic anomalous Hall effect in ferromagnetic metals studied by the multi-d-orbital tight-binding model," Phys. Rev. B **75**, 184416 (2007).

[170] S. Onoda, N. Sugimoto, and N. Nagaosa, "Intrinsic versus extrinsic anomlous Hall effect in ferromagnets," Phys. Rev. Letters **97**, 126602 (2006).

[171] S. Onoda, N. Sugimoto, and N. Nagaosa, "Quantum transport theory of anomalous electric, thermoelectric, and thermal Hall efects in ferromagnets," Phys. Rev. B **77**, 165103 (2008).

[172] H. Kontani, T. Tanaka, D. S. Hirashima, K. Yamada, and J. Inoue, "Giant orbital Hall effect in transition metals: Origin of large spin and anomalous Hall effects," Phys. Rev. Letters **102**, 016601 (2009).

[173] T. Kimura (private communication).

[174] M. Gradhand (private communication).

[175] C. W. Koong, B.-G. Englert, C. Miniatura, and N. Chandrasekhar, "Giant spin Hall conductivity in platinum at room temperature," arXiv:1004.1273v1 (2010).

[176] B. A. Bernevig, T. L. Hughes, and S.-C. Zhang, "Orbitronics: The intrinsic orbital current in p-doped silicon," Phys. Rev. Letters **95**, 066601 (2005).

[177] H. Kontani, T. Tanaka, D. S. Hirashima, K. Yamada, and J. Inoue, "Giant intrinsic spin and orbital Hall effects in Sr_2MO_4 (M=Ru, Rh, Mo)," Phys. Rev. Letters **100**, 096601 (2008).

[178] T. R. McGuire and R. I. Potter, "Anisotropic magnetoresistance in ferromagnetic $3d$ alloys," IEEE Transactions on Magnetics **11**, 1018 (1975).

[179] S. Skalski, M. P. Kawatra, J. A. Mydosh, and J. I. Budnick, "Electrical resistivity of PdFe alloys," Phys. Rev. B **2**, 3613 (1970).

[180] S. Senoussi, I. A. Campbell, and A. Fert, "Evidence for local orbital moments on Ni and Co impurities in Pd," Solid State Commun. **21**, 269 (1977).

[181] Y. Hsu, J. E. Schmidt, M. Gupta, S. Jen, and L. Berger, "Magnetoresistance of Pd-Fe and Pd-Ni-Fe allos," J. Appl. Physics **54**, 1887 (1983).

[182] V. A. Matveev and G. V. Fedorov Fiz. Met. Metalloved. **53**, 34 (1982).

[183] S. U. Jen, T. P. Chen, and S. A. Chang, "Electrical resistivity of Co-Ni-Pd and Co-Pd alloys," J. Appl. Physics **70**, 5831 (1991).

[184] S. U. Jen, "Anisotropic magnetoresistance of Co-Pd alloys," Phys. Rev. B **45**, 9819 (1992).

[185] S. U. Jen, B. L. Chao, and C. C. Liu, "Hall effect of polycrystalline Co-Pd alloys," J. Appl. Physics **76**, 5782 (1994).

[186] J. A. Dreesen and E. M. Pugh, "Hall effect and resistivity of Ni-Pd alloys," Phys. Rev. **120**, 1218 (1960).

[187] H. Ebert and B. L. Györffy, "On the scattering solutions to the Dirac equation for non- spherically-symmetric targets," J. Phys. F: Met. Phys. **18**, 451 (1988).

[188] J. M. Ziman, *Principles of the Theory of Solids* (Cambridge University Press, London, 1972).

[189] I. Mertig, E. Mrosan, and P. Ziesche, *Mutliple Scattering Theory of Point Defects in Metals: Electronic Properties* (Teubner, Berlin, 1987).

[190] P. T. Coleridge, "Impurity resistivities in copper," J. Phys. F: Met. Phys. **15**, 1727 (1985).

[191] M. Gradhand, D. V. Fedorov, P. Zahn, and I. Mertig, "Extrinsic spin Hall effect from first principles," Phys. Rev. Letters **104**, 186403 (2010).

[192] G. Y. Guo and H. Ebert, "Band theoretical investigation of the magneto-optical Kerr effect in Fe and Co multilayers," Phys. Rev. B **51**, 12633 (1995).

Acknowledgements

At first I would like to thank the three main persons from which I have learned so much during my PhD. This is primarily my supervisor Prof. Dr. H. Ebert who gave me the opportunity to work on the fascinating research field spintronics. I am indebted to Dr. D. Ködderitzsch for many fruitful discussions and for his ongoing interest in my work. I want to thank Prof. Dr. J. B. Staunton for her support and inspiring ideas.

I want to thank Dr. J. Minár, M. Kardinal and especially S. Bornemann for their computational support. Many thanks also to Dr. M. Košuth, Dr. S. Chadov, Dr. V. Popescu, Dr. J. Braun and the rest of our working group for valuable discussions and a nice working atmosphere.

I would like to thank my beloved wife Charlotte and my son Maximilian for their patience and appreciation. I want to thank my parents and my brother for their support.

Acknowledgements

Curriculum Vitae

PERSÖNLICHE ANGABEN

- Name: Stephan Lowitzer
- Geburtsdatum: 14.07.1980
- Geburtsort: Dachau
- Familienstand: verheiratet, ein Kind

SCHULBILDUNG

- 1987-1991 Grundschule in Braunfels
- 1991-1997 Gesamtschule in Braunfels
- 1997-2000 Goethe-Gymnasium in Wetzlar

STUDIUM

- 2001 Studium der Mineralogie an der Goethe-Universität in Frankfurt a. M.
- 2004 Aufnahme in die "Studienstiftung des deutschen Volkes"
- 2004 Erhalt des "American Mineralogist Undergraduate Award"
- 2006 Diplomarbeit in der Kristallographie, Titel: "Experimentelle Bestimmung des thermoelastischen Verhaltens von Graphit und theoretische Untersuchung zu Punktdefekten in Albit"
- 2006 Erhalt des Diploms der Mineralogie (Note: 1,0)

BERUF

- seit 2006: wissenschaftlicher Angestellter und Doktorand am Institut für Physikalische Chemie an der Ludwig-Maximilians-Universität München

Curriculum Vitae

List of Publications

- S. Lowitzer, M. Gradhand, D. Ködderitzsch, H. Ebert, D. V. Fedorov, I. Mertig. Spin Hall effect of alloys: Extrinsic and intrinsic contribution. Phys. Rev. Letters, in preparation.

- S. Lowitzer, D. Ködderitzsch, H. Ebert. Coherent description of the intrinsic and extrinsic anomalous Hall effect in disordered alloys on an *ab initio* level. Phys. Rev. Letters, submitted.

- S. Lowitzer, D. Ködderitzsch, H. Ebert. Spin projection and spin current density within relativistic electronic transport calculations. Phys. Rev. B, submitted.

- S. Lowitzer, D. Ködderitzsch, H. Ebert, A. Marmodoro, P. R. Tulip and J. B. Staunton. Why an alloy's residual resistivity can come down when its disorder goes up: an *ab initio* investigation of the K-state. Europhys. Lett., submitted.

- S. Ouardi, G. H. Fecher, B. Balke, X. Kozina, G. Stryganyuk, C. Felser, S. Lowitzer, D. Ködderitzsch, H. Ebert, E. Ikenaga. Electronic transport properties of electron- and hole-doped semiconducting $C1_b$ Heusler compounds: $NiTi_{1-x}M_xSn$ (M=Sc, V). Phys. Rev. B, **82**: 085108, 2010.

- S. Lowitzer, D. Ködderitzsch, H. Ebert and J. B. Staunton. Electronic transport in ferromagnetic alloys and the Slater-Pauling curve. Phys. Rev. B, **79**:115109, 2009.

- P. R. Tulip, J. B. Staunton, S. Lowitzer, D. Ködderitzsch and H. Ebert. Theory of electronic transport in random alloys with short range order: Korringa-Kohn- Rostoker nonlocal coherent potential approximation. Phys. Rev. B, **77**:165116, 2008.